T0291633

BIRD DISPLAY

BIRD DISPLAY

An Introduction to the Study of Bird Psychology

BY

EDWARD A. ARMSTRONG

Omne adeo genus in terris hominumque ferarumque,
et genus aequoreum, pecudes pictaeque volucres,
in furias ignemque ruunt: amor omnibus idem.

Virgil, *Georgic* III.

CAMBRIDGE
AT THE UNIVERSITY PRESS
1942

To

EUNICE

CAMBRIDGE
UNIVERSITY PRESS

University Printing House, Cambridge CB2 8BS, United Kingdom

Cambridge University Press is part of the University of Cambridge.

It furthers the University's mission by disseminating knowledge in the pursuit of
education, learning and research at the highest international levels of excellence.

www.cambridge.org
Information on this title: www.cambridge.org/9781107511576

© Cambridge University Press 1942

First published 1942
First paperback edition 2015

A catalogue record for this publication is available from the British Library

ISBN 978-1-107-51157-6 Paperback

Contents

Illustrations

Preface

AN interesting observation of a bird's behaviour should be no less carefully recorded and reverently preserved than the type specimen of a new subspecies. Lack of regard for this principle has long prevented the outdoor study of birds from being considered much more than the harmless hobby of men who preferred looking at birds to killing them. Now that field ornithology is increasingly recognised to be a serious scientific discipline from which careless observation and wanton generalisation should be sternly excluded, it is essential that its literature should eschew the vagueness which has hampered the progress of bird-behaviour studies in the past. It is not enough to be told that birds do this or that; we should be told what reliable observer has seen them do it. Acting on this principle I have tried to document these pages adequately so that the reader may, if he wishes, trace the source and assess the value of specific observations. If facts are the stones of which the Palace of Science is constructed, sources are the cement. By means of documentation and bibliography I have provided a supply of building materials. No doubt under the scrutiny of architects with larger resources and more spacious plans will rise nobler theories than have suggested themselves to me as I carried my hod of facts and plied my metaphorical shovel and trowel.

Perhaps the belated emergence of field ornithology from the stage of anecdotalism is not surprising in view of the much greater readiness with which man studies inanimate objects and dead structures than living things. Many centuries were necessary before psychology advanced beyond the standpoint of Aristotle; and comparative psychology, of which the study of bird behaviour is not the least important branch, is only just achieving scientific status. The structure of the lifeless bird in the hand is much more completely understood than the behaviour of volatile birds in bushes, but the field ornithologist should be the first to acknowledge the yeoman services of the museum naturalist, as without his prior labours useful field work would be impossible. Moreover, it is now evident that the progress of ornithology is dependent on the close collaboration of the field worker and the laboratory specialist, for the activities of birds are not to be understood without an appreciation of the physiological processes involved. I could wish this book no happier fate than that it might provide both types of worker with some material and facilitate their co-operation.

There must always be a much larger number of people taking delight in watching birds than able to specialise in sustained laboratory or field research. Such naturalists have contributed greatly to our knowledge in the past, and if their efforts are wisely directed they have remarkable opportunities of doing useful, even thrilling, work at the present time. I trust that those called by the pleasant and honourable name 'amateur' will not find as they read these pages that I have cumbered them unduly with technical terms.

The unwary projection of human thoughts, feelings and motives into the minds of animals has vitiated so much scientific work in the past and so many authors consciously or unconsciously exploit a false humanisation of birds and beasts that anthropomorphism has become the *bête noir* of critics, and it is hardly possible for a writer to escape this charge unless he takes refuge in clumsy periphrases and recondite terminology; but if the reader is on his guard, there is no need to write in a style more appropriate to automatons than to birds. If it be true that man never knows how anthropocentric he is, it is not less true that in this age man never knows how mechanistic he is—and a mechanistic philosophy can be no less misleading than anthropomorphism. The discerning reader will not attach a deeper emotional connotation than they deserve to such terms as 'courtship' or 'love' when he finds them in these pages, and he will remember that when I speak of 'Nature' I have in mind Darwin's definition: 'I mean by Nature only the aggregate action and product of many natural laws, and by laws sequences of events as ascertained by us.' Under the term 'Display' I include movements, postures and sounds, generally of a conventionalised kind, which have the capacity to initiate specific responses in other creatures, more particularly in members of the same species.

I have not hesitated to point out similarities between the behaviour of birds and other creatures varying from insects and fish to man. Indeed, one of the outstanding impressions gained as a result of the study of behaviour-patterns is the tendency of certain types of activity to recur in groups of organisms widely separated from each other. These parallelisms are not to be explained as coincidences but are due to the operation of fundamental psychological forces. The ritualisation of activities, for example, characteristic of the human and the subhuman animal alike, is only one amongst many indications of continuity of mental constitution at different levels. Beneath the threshold of distinctively human thought and activity may be perceived processes which find their clearest manifestation in the life of animals. Thus the most profitable technique in comparative psychology is reciprocal, the exploration of the complexities of the human mind aiding the understanding of animal behaviour, the study of man's lowlier fellows revealing mechanisms which operate in the recesses of human personality.

While it is apparent that bird behaviour conforms to definite patterns, only assiduous observation can determine the extent to which variations occur. Future study may show that a few of the instances of display quoted in the following pages arc not entirely typical.

No reader can be more conscious than I am of deficiencies in my treatment of this fascinating theme. There are aspects of the subject with which I have been forced to deal in niggardly fashion; particularly is this true of the relationship between adornment and display. But in order to compress this study into reasonable bounds I have had to restrict the discussion primarily to psychological and behavioural matters. The youthfulness of field ornithology as a science may be best realised by a comparison between the meagre references to bird psychology and display in Alfred Newton's *Dictionary of Birds*, published less than fifty years ago, and the *embarras de richesses* with which the modern student has to contend. He may, however, take courage from the knowledge that there are very few birds whose life history has been at all thoroughly worked out. Great opportunities are his, interesting discoveries assuredly await him. These pages, by chronicling the labours of those who have already borne the burden and heat of the day, may, I hope, be found useful by those who, though later on the scene, may reap a still richer reward.

The numerous references constitute an acknowledgement of the heavy debt which I owe to other writers. My long friendship with Dr W. H. Thorpe has been still further cemented by his patience in reading manuscript and proofs, checking references and providing opportunities for library work at Cambridge. The treatment of the physiological aspects of bird behaviour has profited greatly by the criticisms of Dr F. H. A. Marshall, F.R.S.—who also read the proofs and gave me much encouragement—and the advice of Dr W. S. Bullough with whom I discussed the problems concerned with the relationship between behaviour and internal state. I owe a special debt of thanks to Mr W. B. Alexander, Director of the Edward

Grey Institute of Field Ornithology. He sought out and verified the scientific names of birds mentioned in the text, arranged them in systematic order, placed the resources of the Institute library at my disposal, and in many other kindly ways facilitated my work. Miss Averil Morley also gave much generous assistance in checking references and Mr D. Lack's comments and suggestions have been valuable. I am grateful also to the City Librarian of Leeds and the staff of the Central Library for their efficient assistance. The Leverhulme Trustees greatly aided and encouraged me by granting me an Award. I am very grateful to them. Mr Daniel Lehrman, of New York, sent me an important paper published in Germany since the outbreak of war, and to him and Mr B. W. Tucker I am indebted for some references. Discussion of many a matter with my wife has helped to reduce obscurities of expression and clumsiness of style. Had it not been for the ready help of all these in greater or lesser measure as it was needed, my work, heavily handicapped as it has been by war conditions both as regards observation in the field and research amongst books, would have been conducted amidst difficulties too great for any ingenuity or enthusiasm to surmount.

Dr F. M. Chapman and the authorities of the American Museum of Natural History have willingly allowed me to use some of their photographs. Professor A. A. Allen, of Cornell University, also graciously provided some illustrations, and through the kindness of Mr Lee Crandall and the New York Zoological Society I am able to reproduce a set of photographs depicting the display of birds of paradise and other species. Two friends who were good enough to provide photographs, Miss E. L. Turner and Mr Riley Fortune, have passed away. For other photographs I have to thank Mr George Bird, Dr W. S. Bullough, Mr N. Chaffer, Mr Ralph Chislett, Aircraftsman C. D. Deane, Mr R. T. Littlejohns, Dr L. H. Matthews,

Dr Neal Rankin, Mr Dennis Rankin, Lieutenant-Colonel H. Morrey Salmon, Mr H. B. Savory and Mr Stanton Whitaker. Mr C. Bryant kindly helped me to obtain the photographs of Australian birds. Through the good offices of Professor Spaul some of the illustrations are more adequately reproduced than would otherwise have been the case. It is a pleasant duty to record my thanks to all these.

EDWARD A. ARMSTRONG

Leeds
March 1942

CHAPTER I

THE CEREMONIAL OF THE GANNET

The gannet as a type specimen illustrating ceremonial behaviour — Flying-up ceremonies — Post-nuptial display — Billing and bowing — Vestigial behaviour — The function of tension in social organisation.

IT will be most convenient at the beginning of this discussion to select one species as a representative subject for our purpose—that of analysing bird behaviour and studying the mind of the bird as manifested in display activities. Let us choose the gannet. It is large, conspicuous and readily studied at its nesting haunts, for it breeds in colonies and is not timid; moreover, it indulges in an interesting variety of ceremonies. Early in English history it attracted attention and admiration, for the North Sea was known by the pleasant name, 'The Gannets' Bath', and the author of *The Perils of the Seafarer* wrote,

> At times the swan's song
> I made to me for pastime,
> the ganets cry,
> and the 'hu-ilpe's' note.
>
> *Codex Exoniensis*, tr. B. Thorpe.

To-day about 109,000 of the world's 167,000 gannets breed around our coasts (Fisher, 1940a).

I will ask the reader to imagine himself transported on the swift wings of fancy to the midst of a gannetry. There let us notice some of the birds' curious customs and mannerisms. Later we shall reflect on our observations at leisure, making them the starting point for a discussion of various problems of bird life and the elucidation of their significance.

There is a vitalising thrill in being amongst a multitudinous host of wild creatures; when these are big, elegant and powerful birds in the full vigour of breeding activities the

ABD

1

experience is almost akin to exaltation. The gannets sweep around the stark cliffs, riding gracefully on the wind, alighting with flapping wings and depressed tail to stand ponderously on the rocks; then launching forth again with hoarse *rraah* and sudden downward plunge. In the distance the birds look like a flurry of giant snowflakes, but when we reach the summit of the precipice and look over we see that the ledges right down to the surf-swept rocks are crowded with gannets, for the most part adults, guarding a nestling or incubating; but here and there are immature birds, in their first, second or third year, recognisable by their dark, flecked plumage.

If we investigate the posture of a sitting bird we find that it does not cuddle the egg against brooding patches in the usual way, for it has none, but that the feet cover it. When either male or female settles on the egg, first one foot is placed upon it, the other laid on the first, and then the body is lowered. As a consequence of this peculiar mode of incubation the chalky surfaced egg soon acquires a dirty, yellowish tint. Although this idiosyncrasy of the gannet's was recorded by Gesner (1555) it was long regarded as a myth. It may be compared with the emperor penguin's custom of supporting the egg on its feet—a biological necessity when a bird breeds on ice (Cherry-Garrard, 1922). The Adélie penguin, not quite so austere in its choice of time and place for breeding, lays two eggs and tucks the rearmost on its feet while the other rests on the ground and is covered by the bird's belly (Levick, 1914). I would tentatively put forward the suggestion that the gannet's practice may be correlated with the bird's predisposition to sit on a cliff ledge with tail protruding outwards.[1]

---

[1] So engrained is this tendency that Gurney (1913) found that his captive birds preferred to sit with beaks towards the brick wall of the garden in which they were confined. It is usually maintained that the attitude is assumed in order to trim the feathers to the wind, but this is exactly what the cliffwards position does not achieve. The explanation of the custom may be

The manner in which the gannet leaves the nest is remarkable. The bird rises and stretches neck and beak up to the sky,

of quite a different nature. Nearly all birds take great care not to soil their nests by defecation; and it is essential that they should be particular in the matter because insanitary conditions involve disease and death for the young. Solan geese, brooding their eggs, would jeopardise them and inconvenience themselves if they had to rise and turn round every time they needed to mute; so they sit in a position such that they can squirt the liquid clear of the nest. That the gannet should be fastidious in regard to domestic cleanliness will surprise no one who is acquainted with the strictness with which sanitation is attended to in the bird world. Although roosting doves defecate frequently, an incubating female allows nothing to pass her cloaca throughout the night (Whitman, 1919). A young wren turns round and evacuates at the aperture of the domed nest so that the parent may carry away the capsule, warblers assiduously search for any defilement; in many species the young move towards the edge of the nest in order to void excrement. This is especially noticeable in young crossbills whose parents, by reason of the peculiar shape of their beaks, are unable to remove the excreta. Thus the nest rim becomes deeply encrusted. Montagu harriers have a latrine with a passage leading to it from the nest. Swallows also make runways to the defecation place. Wood-dust sanitation prevents the fouling of young woodpeckers' plumage (Blair and Tucker, 1941). Cormorants and hawks take care to direct the liquid outwards and somewhat upwards. Even before young sparrowhawks can move much they are able to eject their excrement well beyond the nest. Nicholson (1931) describes defecation by a young Guiana king humming-bird as a definite cycle of movements akin to the activities of a young cuckoo in evicting other nestlings. When I was photographing a young brown booby in the West Indies it kept manœuvring to face me, yet although it was in a considerable state of agitation through my presence, when it had occasion to void excrement it turned round in order to direct it over the cliff and not defile its resting-place. Fulmar petrels squatting on ledges before eggs have been laid take care not to defecate towards the rock; but these and other cliff-nesting species—guillemots, razorbills and the like—often drench each other. If young hornbills (*Lophoceros melanoleucos*) can reach the nest-hole they evacuate through it; if not they pick the dung from the floor and put it outside or add it to the plaster of the cavity (Moreau and Moreau, 1940). When a nestling willow warbler failed to deposit the dropping outside the nest the nearest chick picked it up and put it in the proper place (Blair and Tucker, 1941). Such alternative behaviour, not conforming to a rigid pattern, is of great interest. A remarkable adaptation is recorded of the golden-shouldered parakeet; the nest hygiene is performed by lepidopterous larvae (Thomson, 1934).

performing every movement with impressive deliberation. Gazing fixedly at the heavens in an abstracted way it turns, and walking delicately with wings more or less raised and still steadfastly regarding the sky, proceeds to the edge of the cliff and launches itself with a snarling cry.

If we ask, Why does the bird go to all this trouble to take flight? the answer may be given either in physiological or psychological terms. Every bird's body is furnished with airsacs to give buoyancy, but in the gannet they are so extensive as to envelop the breast, belly and sides beneath the skin. Gurney thought that the primary function of these sacs was to break the force of the plunge when the gannet dived; and no doubt they are useful in this capacity both to the gannet and its relative the pelican which fishes in the same way and is also amply furnished with them. Possibly they may also be a protection against extreme cold and in the water they would greatly increase buoyancy and expedite the return to the surface after diving. The advantage of this might well consist in the facilitation of taking flight from the surface of the sea. The gannet has difficulty in rising from still water, but in the ordinary course of events the bird, aided by its air-sacs, would shoot up to the surface with considerable velocity and thus obtain the initial impulse enabling it to open its wings clear of the water. Moreover, the air-sacs may also make it easier for the bird to rise from the ground by allowing of a shift in the centre of gravity when taking off. Possibly these structures serve this purpose in other large birds such as the crested screamer, flamingo, bustard, hornbill and vulture. The crested screamer, for example, though a ponderous bird, spends the greater part of a fine day soaring and singing over the Argentinian plain—as if a turkey were to emulate the lark! But the function of air-sacs is still not fully understood.

Although the gannet's procedure may be of advantage in

PLATE I

page 4

GANNET

Flying-up posture. The position with uplifted wings is somewhat unusual.

pages 58, 190

WANDERING ALBATROSS

The male spreads his wings in courtship.

PLATE II

page 7

GANNETS

Billing ceremony.

page 19

GANNETS

The left-hand bird is eating its nest, the bird on the right is bowing.

securing the partial inflation of the air-sacs there is a formality about it which suggests that it is also in the nature of ritual. The stiffness, deliberation and rapt appearance indicate something more than mere subcutaneous inflation! The gannet's posturings are, in fact, a 'flying-up' ceremony. All large sociable birds make noticeable preparations when about to take wing, and some of these initiating movements have no apparent usefulness so far as rising from the ground or water is concerned. The mallard crouches and jerks up head, neck, and forepart of the body in a way similar to, but slightly different from, the actual motions of leaping upwards; the Egyptian goose shoots its head into a rigid vertical position; the grey lag goose holds its neck stiffly in the air, at the same time shaking its head and beak side-ways; the whooper swan adopts an up and down motion of the head. The utility of these movements in signalling forthcoming action is obvious. It is of great advantage to birds which migrate in flocks, such as geese, to take flight so far as possible simul-taneously, and thus range themselves without delay into orderly squadrons. Moreover, the movements serve as a quiet hint of danger to neighbours when a bird sights a suspicious object. They have, in fact, a contagious effect. Large gaggles of geese in which one or other of the birds is constantly initiating flight in this way fly up much more often than small parties. The posturings of the solan goose in all probability have a somewhat different significance from the ceremonies of the true geese. It is rather clumsy both in taking off and alighting, and liable acci-dentally to interfere with birds close by, so it shows its inten-tions by this accepted code and thus avoids misunderstanding with its rather querulous neighbours, minimising the possibility of a vicious stab at the critical moment of taking off, and re-ducing the danger of injury to its wings by contact with birds around it, or collision with those returning.

It has been worth while to devote a little time at the beginning of our discussion to these details because we are thus reminded that the activities of birds should be considered as behaviour-patterns in which each element is related to others. We have seen that the 'flying-up' ceremony of the gannet must be studied in the light of the bird's structure, its social relationships and its method of incubation. As our survey proceeds we shall do well to remember that to bring order into our complex material it is constantly necessary to scrutinise certain activities out of their usual living context; unless we keep in mind the artificiality of this method, deplorable but unavoidable, we may gain a false impression of the significance of the particular kinds of behaviour which we are studying.

On my first visit to a great gannetry in August, when nearly every pair of gannets was tending a chick, I anticipated that almost their entire attention would be concentrated on domestic duties; but, on the contrary, I found that they were still in the mood to give prolonged ceremonial displays. Of course these posturings cannot be regarded as courtship blandishments, for indeed the evidence indicates that gannets arrive paired at their nesting sites and may even mate for life. We shall do well to regard this first encounter with post-nuptial display as a pre-liminary monition that we must consider the term 'courtship' when used of birds as having a different connotation from that which it usually bears. Perhaps 'courtship' should be restricted to pre-nuptial display and some such term as 'connubial display' reserved for post-nuptial demonstrations, but in practice it is awkward to keep to such strict terminology, especially as the same display may occur in different phases of the breeding cycle and vary in its significance. It is important to remember, however, that pairing up may take place with little or no display, as in the case of robins (Lack, 1939a).

When a bird returns to its mate, or as the pair stand together on a ledge, they face each other, stretch up their heads with beaks pointing skywards, partially open their wings and waggle their heads energetically so that the beaks clatter as in a kind of fencing, or scrape together as if they were being whetted one upon the other. At this point one of the birds may open its mouth capaciously showing the wide, black gape (Massingham, 1931). The clashing of the big bills reminded me of the 'klapper' which a stork makes with its mandibles as it stands on the nest. The gannets' ceremony is accompanied by a hoarse, throaty shout—*rrah, rrah, rrrraaaah*—which is often taken up along the ledges in a strident, querulous chorus, so that at a little distance there is a strong resemblance to the vociferous remonstrances of a football team's supporters when a doubtful offside has been awarded.

Billing of one kind or another is a very frequent element in bird courtship. The male puffin approaches the female on a rock or grassy patch, and as a prelude, nibbles at her bill; then they stand breast to breast shaking their heads rapidly so that the horny plates clash, and in the midst of these doings they bow to one another (Lockley, 1934). Ivory-billed woodpeckers— now, alas, great rarities—clasp each others beaks (Allen and Kellogg, 1937). There is a kind of kissing or 'nebbing' amongst birds as various as the great crested grebe (Selous, 1933), heron (Selous, 1927), hawfinch (Nicholson, 1927), black guillemot (Armstrong, 1940), parakeet, and cedar waxwing. The male of the last-mentioned will place his bill in his mate's mouth (Crouch, 1936). Amorous ravens hold each other's bills in a prolonged 'kiss' (Coward, 1920), and a female thrush has been observed to peck into the open mouth of the male after the nuptial rite (Ellison, 1931). It may also be mentioned that courting elephants place the tips of their trunks in each other's mouths. Darwin (1845) describes and Crawley (1902) compares with animal behaviour the ceremony of rubbing noses and similar customs amongst human beings. Rothmann and

Teuber (1915) believe that in chimpanzees kissing and the transference of food from mouth to mouth are connected, a suggestion which should be borne in mind when we presently discuss courtship-feeding amongst birds.

What purpose can the gannets' ceremony serve? Courtship days are past, family responsibilities press upon the birds, and yet these vigorous endearments continue. Indeed, they go on all through the breeding season and have been noticed as late in the year as 19 September (Gurney, 1913) and as early as January (Robinson, 1935). They are thus to be reckoned connubialities rather than wooings. We shall defer this problem for later consideration, only pausing now to draw attention to the fact that the species in which this type of belated mutual connubial activity occurs are those in which the plumage of the two sexes is alike, or nearly so, such as the great crested grebe, the red-throated diver, the gannet, and albatrosses. Moreover, it is most noticeable in birds which are not classified near the top of the avian tree.

Often associated with the gannet's 'fencing', 'scissoring' or 'billing' is another connubial 'figure'—to borrow a term from folk-dancing—which may be called 'bowing'. As the bird stands on a rock the neck is arched and the head brought swiftly along the flank as far as the foot, or even slightly beyond, giving the effect of a hurried, low obeisance. Simultaneously the wings are spread and waved, the tail raised and lowered. The gannet will sometimes conclude the performance with a contented waggle of its tail like a duck emerging from a pond. In its most complete form both birds take part in this display, but it is often performed by a single bird which makes its salaam to nothing in particular. When the display is mutual it becomes a rhythmic duet. As I watched the birds I was reminded vividly of a scene frequently to be observed in a Japanese street—two gentlemen from the country engaged in a polite colloquy, consisting of vocal honorifics interlarded amongst almost interminable bowings, while the life of the street flows on unconcernedly around them. The solo performance is executed regardless of whether the bird's mate is in the neighbourhood or not; he—probably I could with equal accuracy write she—stands on a rock or ledge and bows when the spirit moves him, entirely for his own satisfaction apparently, as his neighbours take no more interest in his doings than the citizens of a Japanese city of the courteous kotows of strangers. These antics, too, go on all through the breeding season. Solo displays, it may be mentioned, are not infrequent amongst birds. A caged dove will posture alone and solitary terns will often do so (Marples and Marples, 1934). Isolated jumping spiders, too, sometimes perform sexual dancing (Berland, 1927), and at the other end of the scale chimpanzees will dance a *pas seul* (Yerkes and Yerkes, 1925).

The post-nuptial 'fencing' and 'bowing' are not the only survivals or continuations of an earlier psychological phase to be observed in a gannet colony. In mid-August as I sat on the edge of a cliff watching the birds flying past, arriving and departing, quite a number were to be seen carrying nesting material, varying from withered grass to wet seaweed. Moreover, I noticed several gannets stealing material on the borders of their fellows' nests. This was not done in a business-like way; there was no crafty arrival, rapid, furtive snatching of plunder and hurried departure, such as is characteristic of thieves in a gullery (Kirkman, 1937) or penguin rookery (Levick, 1914). The birds came, picked up a few stems or trailers of weed, held them in their bills, looked about for a while, let them fall, picked up more, dropped most of this, and so on. These observations were made at a season when one would have expected nest-building —such as it is in gannet circles—to have been completed, but on 15 September, when Gordon (1921) was on the Bass Rock, birds were assiduously collecting material of all sorts for their nests, although sixty per cent of the young gannets had already flown. Even in Christmas week the birds have been seen carrying stuff about (Robinson, 1935). Not only did the robbers conduct themselves in the most nonchalant way but there was no outcry raised, no open disfavour expressed, no blows struck in defence of the owner's property, none of that corporate moral indignation which writers in the past believed to be characteristic of such colonies of birds. There may be corporate action, but it certainly has not a moral basis. Would-be poets may sing of the rooks:

> When they build, if one steal, so great is their zeal
> For justice that all at a pinch
> Without legal test will demolish his nest,
> And such is the trial by Lynch.
>
> W. J. Courthope, *The Paradise of Birds.*

but those who have studied these birds are unable to confirm this popular idea that thieves are mobbed (Yeates, 1934). F. M. Chapman (1908a) records that brown pelicans 'steal one another's nesting material with an air which plainly bespeaks a knowledge of their guilt and that they expect punishment from the birds they have robbed', but this never leads to more than a bloodless squabble between the individuals concerned. Robber penguins are chased and punished by their victims if they can catch them (Levick, 1914; Roberts, 1940a); black-headed gulls manifest community animosity against marauders up to a certain point, but that this is not due to altruism or public-spiritedness is sufficiently shown by the conduct of a gull, which after driving off a thief returned to its nest with a supply of material from the nest it had defended (Kirkman, 1937). Earlier in the season, however, according to some writers, petty larceny amongst gannets is more purposeful, and if the robber is de-tected he or she is assailed with stabs by the victim. A nest is seldom, if ever, left unguarded, so the pilfering is commonly done slyly by a bird with neck outstretched to a neighbour's nest.

It may be that a certain amount of this playing with material which I noticed was due to the gannet's nervousness at the presence of a spectator, much as a highly strung man will fidget with his hat or watch-chain, or a preacher develop some manual mannerism. Chapman (1908 b) found the boobies on Cap Verde, both young and old, had this habit, which he attributed to the trepidation caused by his proximity. This is consonant with the view shortly to be advanced that playing with nesting material and belated nest-building are expressions of side-tracked emotion or due to thwarted activity. Fidgeting with weeds, sticks, grass and so forth, goes on even when the birds are not aware that they are under observation, though, of course, nervousness may accentuate it. There are colonies of boobies, tropic birds and frigate birds on Melville Island off the coast of Tobago. I spent some time concealed in the upper branches of a small tree in which frigate birds were nesting at a time when these, as well as the boobies, had well-grown young, and watched boobies stealing sticks in a leisurely and apparently purposeless way from the frigate birds' nests. Noddy terns, in common with gannets and boobies, delight in carrying material about during the whole of the breeding season. As early as their sixteenth day the young act thus (Watson, 1908). Skimmers do likewise when a fortnight old (Pettingill, 1937). Precocious behaviour of this type considered in conjunction with such in-stances as the rearing up, fencing and weed-dangling of young great crested grebes exactly as adults do when courting (Coward, 1923, 1926), apparent coition behaviour in Bicknell's thrush when only a month old (Wallace, 1939), the brooding of its mother's egg by a dove twenty-one days old (Craig, 1909), and the singing of young red oven birds in the nest (Hudson, 1892), indicates that we should be careful how we use the words 'sexual behaviour'. It seems clear that certain actions are without direct sexual reference until the organism has reached the phase in which they function in due sequence in the whole sexual pattern. Thus, if the Freudian psychologist wishes to use these facts in support of his theories he must first show that such types of behaviour can properly be regarded as having an erotic element. Young black-headed gulls artificially reared together associate in pairs at the age of six weeks (Rothschild, 1941). Is this or is it not a manifestation of sexuality?

Leaving out of account the extreme nonchalance of the gannets and the fact that at the time when I studied them they seemed merely to be toying with the material and had apparently no definite object in their thievish practices, and granting that some, however few, did add their garnerings to their nests, what purpose could all this activity serve? It may be argued, and indeed has been by Gurney, that additional material is needed, owing to the structure being flattened by the growing chick; also, the wind may blow portions of it away from time to time as it dries. But the size of the nest varies immensely—from five feet in height to a thin layer of seaweed—and there is nothing to indicate that a youngster in an ample nest thrives any better than his neighbour in a cradle of meagre dimensions; nor is there any evidence that the object of the addition of material is to build up the nest to prevent the young bird from falling over its narrow ledge. The nest, in actual fact, provides little protection of this kind, and the chicks are so adapted to their narrow environment and precarious situation that they move but little. Still less may we suppose that the birds prefer new, dry material, for wet seaweed is the basic substance of the nests. The *Fucus*, *Laminaria* and *Ascophyllum* used adhere to the rock as they dry, and it is very seldom that a nest is blown away. Gordon (1921), indeed, speaks of a gannet rejecting straw and choosing seaweed, but his interpretation of the incident, which he calls 'a clear case of reasoning', is unsound. He refers to a bird returning to its nest with straw and seaweed and departing with the straw on finding that its mate had not incorporated a previous offering of straw into the nest. The birds in general are too careless with the material which they pick up for it to be claimed that reasoning of any sort enters into this situation. 'Reasoning', indeed, is a term which is best avoided in speaking of birds. The gannet which plucked feathers from its own plumage and placed them carefully in the nest might have been considered specially intelligent had it not, a few moments later, flapped its wings and heedlessly blown them away (Booth, 1881–7). Gannets do, in fact, add straw to their nests whenever it is available, and have been known to appropriate such things as a red coat, a brass sun-dial, an arrow presumed to be of Indian or Eskimo origin (Martin, 1703), golf balls, small branches, blue castor-oil bottles, strings of onions and a clockwork steamer.

The blind uselessness of vestigial nesting habits so far as the young birds are concerned is illustrated by an incident described by Gordon (1921); just after its offspring had flown the parent was seen to be rearranging the nest, paying no attention to the efforts of the youngster in its first excursion into the wider world. Equally incongruous was the behaviour of the cormorant which Perry (1940) watched pilfering material from its own nest.

Vestigial habits I would define as behaviour-patterns which survive beyond their period of usefulness. They are of two kinds: racial and cyclic. Roosting penguins are able to place only the tips of their beaks under their apologies for wings, yet they continue a practice which served a useful purpose when their drowsy ancestors put their heads under ampler wings. This is an example of a racial vestigial custom. Similarly the sailor's collar, the buttons at the back of a tail-coat, and the archdeacon's gaiters remind us of the days when seamen wore pigtails, gentlemen carried swords, and church dignitaries made their visitations on horseback. When birds continue habits beyond the stage in the reproductive or life cycle in which they serve a useful function we have cyclic vestigial customs. They may be characteristic of the species, as in the case of the belated nesting activities of the gannet, or only occasionally noticeable in particular birds. A snow bunting engaged in feeding young will fly back to the territory which they have vacated to defend it needlessly against another male (Tinbergen, 1939*a*). Guillemots, having lost a pipped egg or young chick return to the nest site and perform the food-offering ceremony for two or three days after their loss (Johnson, 1941). A dotterel, as we shall see later, will perform the 'broken-wing trick' when the young have flown out of danger. There is hardly a behaviour-pattern which may not thus outlive its usefulness. Cyclic vestigial behaviour may be partly due to the tyranny of habit, but unquestionably its main cause is a lack of correlation between the bird's internal state and the environment. This is also the explanation of types of precocious development such as were mentioned a little earlier in which activities properly associated with the reproductive cycle appear disintegrated from the whole appropriate pattern of behaviour.

The gannet, like some favoured human beings, has emancipated itself from the worst rigours of the struggle for existence. Its enemies are few, its food supply plentiful. Without endangering the race it may work off surplus energy in harmless posturing. Dance and song, as we shall see, though subserving important biological functions amongst birds, are often carried to such a pitch that they seem to be performed, as amongst human beings, for their own sake and in some sense for the enjoyment derived from them. Although the gannet's antics may be attributed to the stimulation of the endocrine system, the psychological aspect is no less important than the physiological.

As we watch a gannet wheeling in the sky and then plunging boldly into the sea we might think the bird a perfect symbol of unrestricted freedom. And yet a little investigation at close quarters shows that its liberty is restricted in countless ways by what we may style the bonds of its own personality, and the 'conventions' of society. Like the apparently free and unrestrained Australian aborigine the gannet is dominated by a most rigid social system. It makes love and war ceremonially; there is a ritual of taking flight, of greeting, of nest-building and of feeding the chick. All this would not be so unless in some measure ceremonial brought happiness—or at least a sense of well-being—and prosperity, to the race.

Man may learn of the gannet as well as of the ant, and human society would be happier if modern trends were less towards the social organisation of insects and more towards the community life of birds. Out of the stresses in the individual as well as in society the gannet has evolved a variety of stately ceremonies which add interest and loveliness to life. Tension is the source of progress and the forcing ground of art, and it is when individuals, or the society which they constitute, timorously and weakly abandon their duty to make this tension constructive that disaster ensues.

CHAPTER II

THE EVOLUTION OF NEST-BUILDING

Widespread pseudo nest-building habits amongst birds—Fidgeting with nest material in nuptial activities—Destruction of nest or eggs due to emotional stress—Relationship between physiological processes and nest-building—Courtship activities associated with nest-construction—The origin of nest-building from fidgeting with material during emotional excitement.

THE preceding description of the ways of nesting gannets has shown that ceremonial plays a very important part in their lives. The gannet is not exceptional in this respect though it exhibits a particularly wide range and variety of display patterns. Let us now, in this and the immediately succeeding chapters, consider some of these forms of display in detail. We shall find that most of the gannet's queer antics have analogies elsewhere in the bird world, and that a survey of them gives us a glimpse into that realm of which we, as yet, know so little, the mentality of birds.

The gannet, as we have noticed, manifests a proclivity for carrying about straw and seaweed long after these things could be of use for nest-making. The nature of this kind of reaction is clearly shown by Kirkman (1937) in his study of black-headed gulls. When a broody gull is prevented by its mate, or otherwise, from sitting on the nest, it will commonly go off to fetch material, so that the size of a gull's nest is largely determined by the number of times it is hindered from incubating. Thus some of the nests become absurdly big, and the pile, not being an integral part of the nest, becomes merely a convenient storehouse from which pilferers may help themselves. As in the case

PLATE III

page 9

GANNET

Solo bowing ceremony.

pages 9, 16

GANNET

Vestigial nest-building.

PLATE IV

page 17

HERON

Stick collecting continues although the young are a month old.

page 26

ARCTIC TERNS

Scrape ceremony. The male works around the female.

of the gannet the addition of material goes on even after the young have hatched. I have seen a coot make fourteen journeys to the nest with weeds in ten minutes, although the structure was already large. It transferred them for manipulation into the nest to its incubating mate, and after the series of journeys was completed took its partner's place on the eggs. Herring gulls, little gulls (Goethe, 1937) and common terns (Steinbacher, 1931) bring nesting material when they arrive for their shift on the eggs; these terns line their nests after they have begun to sit (Tinbergen, 1935). Rooks, storks, herons, eagles, buzzards and hawks bring sticks or branches to the nest when the young are well grown. The raptors prefer green branches—for instance, the white-bellied sea eagle of the China coast sometimes partially covers the eggs with leaves of *Rhodomyrtus* (Aylmer, 1932). Buxton (1932) describes as 'a ceremony' the bringing of sprigs to a honey buzzard's nest and their subsequent manipulation. Every morning the hen brought two branches and mother and chick played happily with them. Brush turkeys go on scratching up leaves into the heap in which the eggs are buried for months after they are needed (Thomas, 1939), and a cob mute swan continues sweeping up weeds for the nest when they cannot be of any use (Perry, 1938).

The attraction of odd or pretty objects for nesting birds is notorious. Autolycus (*The Winter's Tale*, IV, iii) advises the housewife, 'When the kite builds look to lesser linen'. A pair of Bombay crows stole £25 worth of gold spectacle frames from a workshop window (Dewar, 1928). Snake-skins decorate the nests of many species (Finn, 1919), and the satin bower bird concentrates on collecting blue objects for its bower (Chisholm, 1934). Other instances of nest adornment, such as the flower gathering typical of starlings and the taste for shells characteristic of gulls (Goethe, 1937), are too well known to need further reference.

Birds may fidget with material in a great variety of situations. Terns and Waders draw blades of grass through their mandibles as they incubate. Sitting on the eggs a merlin will seize twigs in its bill, at the same time moving the anal region like the male *in coitu* (Selous, 1913–16). Peacocks scratch the ground and collect straws between bouts of displaying, and the grey peacock pheasant will throw a seed or pebble at the female during his display. Gannets pick up material in the intervals between bowing and fencing; they will even sit on the nest site, act as if taking material from an 'imaginary' mate and build it into a fictitious nest (Lorenz, 1937). Boobies, both old and young, vent their feelings by picking up bits of stick and grass (Chapman, 1908*b*). Angry avocets throw straws and shells about (Makkink, 1936). The Adélie penguin presents a small stone or some snow as a token of his regard (Levick, 1914; Ponting, 1921; Roberts, 1940*a*), and the female skua drops grass before the male in courtship (Perry, 1938). During their nuptial activities grasshopper warblers carry nesting material about; little grebes and great crested grebes dive for weeds and dangle them in their beaks, the former in a rather haphazard way, the latter in a definite ceremony (Huxley, 1914). Even the guillemot scratches with its feet in courtship and presents nesting material to its 'intended' (Johnson, 1941). Black-headed gulls will break twigs from tall trees in spring (Portielje, 1928). A female cuckoo has been seen holding a twig or straw in her bill for over a minute while being wooed by the male (Chance, 1922). This may be a vestigial action harking back to the days when cuckoos built nests. A male black skimmer walks up to the female with a stick; she approaches with a coy demeanour and suddenly seizes it with an upward jerk. The nuptial rite follows (Pettingill, 1937). Before nesting the ruffed grouse throws leaves over her shoulder (Allen, 1934). A red-throated diver, squatting where her nest will ultimately be, casts grass and moss behind her after copulating (Selous, 1927), and many other birds pick up leaves, grass, pebbles, or twigs during the sexual frenzy. The female southern cormorant pulls at twigs after assuming her extraordinary soliciting posture, during which she erects her tail and twists her head so that the crown nearly touches her back. After coition she resumes this attitude and the male picks up a stick or other material and offers it to her (Portielje, 1927). A large number of species, especially amongst the Waders, cast fragments of herbage behind them during the 'change-over' at the nest. To the list given by Tinbergen (1935) may be added the curlew (Gordon, 1936), black-winged stilt (Strijbos, 1935), and common screamer (Stonor, 1939).

I have seen a gannet eat part of its nest. As I was only a few feet from the bird, I was able to observe it carefully and to photograph it in the act. It picked up bits of weed, stem and dirt from the nest, and, holding its beak on high, swallowed them with violent and vividly perceptible movements of the gullet. Not once or twice, but again and again, many times in succession did the bird do this. Probably this behaviour was the outcome of nervous agitation caused by my presence. Perry (1940) records having watched a razorbill pick up and swallow 'little bits of dirt and stuff' when his mate pushed him away and took his place on the egg. We shall see later that ceremonial frequently accompanies the 'change-over'—an indication that it is an occasion of nervous strain.

I have watched a raven, anxious for its young when I was near the nest, tearing up bunches of grass to send them floating on the wind. Darwin (1845) describes how chimangos tear up grass in rage. A cock herring gull when excited by intruding males will pluck the grass beside him and ram it into a hollow with his breast (Howard, 1935). According to Goethe (1937) the gull's plucking of herbage is a symbolical form of threat-fighting analogous to the posturing of the ruff. During territory disputes, and possibly also during pair-formation, marsh tits pluck moss from trees (Morley, personal communication). A sarus crane has been seen to scatter the material of her nest furiously when she found that her egg had been taken (Hume, 1890), and there is circumstantial evidence that male crimson-crowned bishop birds destroy their nests, if robbed, and carry away the materials (Lack, 1935). Richard Kearton (1903) watched a ring ousel tear the lining of her nest to pieces after he had moved it a little way from its original site. On the other hand, wrens of the species *Heleodytes nuchalis* packed the entrance to their nest with silk-cotton and feathers while a collector was engaged in cutting off the branch on which it rested (Cherrie, 1916).

Even the eggs may be destroyed by an agitated bird. Selous (1905a) saw a waterhen go to its nest, transfix and carry away five of its eggs in succession. It deposited each egg in the water. An American ornithologist found the nest of a Virginia rail with nine eggs but did not disturb it. When passing again five minutes later he saw a rail, which he presumed to be the female, stabbing the eggs with her bill; three had already been destroyed (Bowles, 1925). In connexion with Swainson's motmot Belcher and Smooker (1936) record that they found a bird sitting on addled eggs: 'On being disturbed she backed off them to the end of the chamber, but when an egg was withdrawn a little by means of an improvised rake she rushed down the tunnel after it and stuck her beak into it.' Turner (1924) saw and photographed a nervous water rail removing egg and young from the nest. The removal of their eggs by nightjars is fully substantiated (Audubon, 1840-4; Jackson, 1938; Bent, 1940). Before deserting their colonies Sandwich terns (*Report of Wild Birds Protection Committee of Norfolk Naturalists' Trust*, 1938) and American tri-coloured redwings (Lack and Emlen, 1939) apparently pierce their own eggs. Caspian terns when intruded upon at their nesting colony hovered over their nests and struck the eggs with their beaks, breaking at least a quarter of the eggs in the colony (Aymar, 1935). Wilson (1907) watched a perturbed Adélie penguin demolish its egg.

In some instances toying with odds and ends may be regarded as an inherited tendency carried over from generation to generation as in the case of young skimmers, sometimes manifesting itself as a vestigial habit, as when the gannet collects weeds in September or busies itself rearranging the nest just after the young bird has flown. But many of these practices may be regarded as providing emotional outlets. They are substitute activities, common enough in human life. An illustration may be culled from Sir R. Storrs' autobiography *Orientations*. When his father accidentally broke some valuable china his mother smashed a ball-room chair to matchwood and said 'almost without emotion': 'There but for the grace of God goes John Storrs.' In grief, chagrin or rage man 'rends his garments', an angry bull gores the matador's discarded cloak, an elephant pushes down saplings, and even a fish will ravage the aquarium. When a tigress, disturbed in her lair, eats her cubs her action may be attributed to nervous stress similar to that which caused the gannet to eat its nest or the terns to destroy their eggs. When crossed in love a man may buy a cornet; and a cat, unable to escape from the psychologist's puzzle-boxes, will wash itself with exaggerated care (Adams, 1931). As we shall see later, the preening, flapping of wings and bathing which in many species of birds accompany display are more than matters of toilet; they are means of emotional relief and expression. When an emotion is thwarted or an activity is denied its natural outlet, relief is found in some other direction, sometimes useful or potentially useful, often useless or destructive. If this fact were more fully recognised by politicians and educationalists, the world would be a happier place.

These examples of abnormal behaviour emphasise the close

connexion between emotional states and the picking up and disposal of material in the immediate neighbourhood of the bird. They show that the normal manipulation of such material —nest-building—is the response to an emotional urge or drive.

The gannet's dilettante fidgeting with nesting material, however out of place in August or September, is characteristic of an earlier phase in the breeding cycle of many birds, as the courtship actions referred to serve to show. A male bunting or warbler will trifle with sprigs or stems and even build a nest which the female may eventually use for her eggs; and she may toy with stems and leaves, neb them and let them fall (Howard, 1929). A wren or a waterhen builds 'cock's' nests. 'Display building' is characteristic of the dabchick and, indeed, of many other birds. A night heron will pull lackadaisically at twigs early in the season without definitely trying to build (Lorenz, 1937). If a bullfinch first constructs an incomplete and amateurish nest and then a perfect and typical example, the imperfections of the first are not due to lack of experience; rather, at the first attempt the bird's internal state was not properly coordinated with external conditions (Morgan, 1930). When a bird is not physiologically mature for the activities of a particular phase in its cycle the execution of the reactions appropriate to that phase will be incomplete or abortive. It is often in such circumstances that birds fidget aimlessly with material or build imperfect nests. When physiologically ripe for the nest-building phase a bird builds with address and efficiency, but even an experienced female will construct abortive 'shells' and platforms when anticipating her second brood if her organism is not correctly adjusted to the situation. There is no evidence that faulty technique is due to the individual's inexperience; the bird carries within it the capacity for building nests after its kind. Marais (1937) reared four generations of weaverbirds without giving them the opportunity to see even a scrap of nesting material, but their descendants plaited their elaborate nests perfectly. Verlaine (1934), however, experimenting with canaries descended from countless generations reared in artificial nests, found that when supplied with suitable materials the first nests were shapeless and took a long time to complete, but succeeding nests rapidly approached the normal and were quickly constructed. Birds reared in these were somewhat superior in nest-building to birds reared in artificial nests. Verlaine concluded that such birds remembered their nests. Further experimentation is desirable as it is more probable that the increase in efficiency was due to better co-ordination than to any process of memory.

Thus, before and after the actual construction of the nest pseudo nest-building activities may take place. Although divorced from their normal, technical end, they are not necessarily useless, for before the nest is built they may play an important part in the courtship ritual, and afterwards they provide a wholesome or harmless outlet for emotion. Even possessing such a safety valve the gannet sometimes courts his mate with such ardour as to denude part of her neck of feathers!

In the ordinary way, as Howard (1929) shows, 'the course of sexual activity and the course of building activity not only run along parallel lines but the changes in each synchronise', and effective building is an indication that copulation has taken place. Either male or female may toy with material in the earlier phases of the reproductive cycle, but whereas during the male's first phase singing, fighting, territory-seeking and general demonstrativeness may be contemporaneous the female's reproductive cycle appears to be under closer physiological control. At one moment the male will chase the female with a sprig or a leaf in his bill, at another he weaves it into a nest; but the female's activities are not so diffused. She may indeed fidget with material before she is ready to receive the male, but once this stage has been reached her nest-building goes apace, first the outer shell and then, in due course, the lining. It is impossible to resist the conclusion that nest-building is correlated with a precise physiological state, especially as females late in finding mates not only go through the early sexual stages swiftly but build without repetition or pauses. Ovulation and nest-construction are undoubtedly closely connected. The female commonly shows her readiness for the male by specific attitudes, activities or calls, so that posturing, too, is certainly correlated with the physiological condition of the bird. As this discussion proceeds the close association between display and the activity of the endocrine system will be increasingly evident.

Belated, aimless or redundant sexual activities, such as display and the carrying about of nesting material, may sometimes be considered as indications of thwarted or incipient double-broodedness. After a red-necked phalarope had laid her last egg she began ceremonial flights once more (Tinbergen, 1935). Other species show a recrudescence of display after feeding the young is completed. In favourable circumstances a second brood may result. Some insectivorous birds, such as swallows, by reason of the comparatively rapid fledging of the chicks, are able to rear two, or even three, broods in a season, but this is quite impossible in the case of storks and gannets whose young take longer to rear. Thus their surplus energy dissipates itself in display activities. The fact that blackcaps, for example, are more often double-brooded in the south than in the north of England indicates how delicately correlated are environment and the endocrine balance. Presumably there is some kind of physiological control which tends to prevent fruitless attempts to rear second broods; but this field of investigation still awaits exploration.

The birds which build the most elaborate nests add little or nothing to them once they are completed. The nest-building impulse works out to its consummation and lapses when the eggs are laid. It is a phase of the reproductive cycle which is superseded when it has run its course and the nest is ready for the eggs. But with such species as gulls, gannets, hawks and storks, we have the raw material out of which artistry is evolved rather than artistry itself. There is a tendency for birds whose nest-building is technically of high quality to reach a culminating peak of achievement which exhausts the motive power behind the impulse and leaves free scope for the brooding and feeding activities. In the case of the less 'artistic' species which merely pile material together, the phase, like the nest itself, is rather amorphous and peters out rather than finds completion. The delimitation of the sphere and term of other sexual activities is equally indefinite in the case of such species as the gannet. This lack of specialisation is not always a disadvantage or an indication of stupidity—though the brain capacity of the gannet is not great. It may be that the ability is but dormant and that the raw mental material still awaits exploitation. So long as a bird has an unspecialised technique there may be latent the possibility of developing it. Too precise specialisation in a changing world may be a greater danger than too little, for the specialised organism may have to deal with the problem of unlinking the sequence of its behaviour, and it is never easy to blaze trails away from the beaten and accustomed track. The great auk and dodo might be alive to-day had they not specialised themselves to such a high degree, the one to an aquatic, the other to a completely terrestrial existence.

Nest-building we have seen to be a response to an emotional drive and we are now in a position to realise how it arose. So close is the connexion between nuptial display and nest-construction that Selous (1901–2, 1905 a, 1916, 1927, 1933) has been able to produce a formidable array of facts to support his thesis that the origin of nest-building is to be found in the movements of the birds in sexual frenzy during copulation. One aspect of this is illustrated by the fidgeting of red-throated divers with grass and stems after the consummation; another by the fact that the female snow bunting's nesting activities are most persistent after coition (Tinbergen, 1939a); yet another is evident in the making of a 'scrape' in the soil or sand as an element in the courtship ceremonial. A month before herring gulls lay their eggs they treat the square foot of ground on which these will be deposited as the focal point where they express emotion in a display which has the effect of wearing a slight hollow in the ground which will later become the nest (Darling, 1938). These movements may take place on roofs or, in the case of the common gull as well as the herring gull, on the water (Portielje, 1926; Goethe, 1937). When a fish-bearing tern approaches a female she will commonly go into the attitude of supplication, crouching with body low, head and tail raised and wings half-open. If the male now walks round her, as he often does, she turns to face him and thus scoops out the nest-hollow with her breast and feet (Coward, 1923). Marples and Marples (1934) noticed that a male tern will posture, and then make a scrape which is occupied by the female while he circles and she turns to face him, thus enlarging the hollow. Southern and Venables (1936) think that the female great skua does this to prevent the male mounting, but it helps to form a receptacle for the eggs. The male lapwing picks up pieces of grass and revolves on his breast before the female, the while keeping his banded tail elevated and expanded and the chestnut under tail-coverts exposed (Coward, 1923). The cock ringed plover makes scrapes, uttering a triple note while doing so. All the time he keeps picking up minute pebbles and with head thrown back lets them fall in a little stream around the hollow (Turner, 1928). Black-tailed godwits (Huxley and Montague, 1926), stone curlews

(Portielje, 1928), ostriches (Selous, 1901), nightjars (Heinroth, 1909) and red grouse (Nethersole-Thompson, 1939) also perform a 'scrape-ceremony'.

An unmated female red-necked phalarope makes scrapes in the herbage, and from the first day of finding a mate this 'ceremony' often follows coition. Twice or thrice a day there is joint scraping. Before laying, the female visits the various scrapes and lays an egg in one of them. She pecks the ground and the male follows suit. He then walks away throwing scraps of vegetation over his shoulder. Back he comes, stimulating his mate to renewed pecking. Thus it continues some twenty times. Finally, the female departs and the male sits on the egg and begins to build, picking up sedge stems and placing them behind his back in the nest on one side (Tinbergen, 1935). Thus scraping and picking develop into nest-building and are stimulated emotionally. It is highly significant that the places where emotion was expressed by scraping should be remembered. When the migratory bird returns to the neighbourhood which it tenanted the previous year we have a somewhat similar phenomenon on a much more impressive scale. Joint nesting activities of a more or less symbolical kind unquestionably have a stimulating effect on grey herons (Verwey, 1930) and terns as well as red-necked phalaropes.

Thus we find the collection and shaping of material represented in the nuptial antics and can conceive how the present immense range in nest patterns evolved from these crude beginnings.

Not only does nest-construction seem to have arisen from fidgeting consequent on emotional stress, but also, possibly, the practice of covering the eggs on leaving them. Ducks denude their breasts of down to make a concealing blanket for their eggs; perhaps, originally, the plucking was due to emotional tension. In a Yorkshire wood I find that mallards have learned to pile large heaps of leaves over the clutch. Partridges also use grass and leaves to cover their eggs. From such beginnings it is easy to imagine how brush turkeys learned to construct their natural incubators. But we must be on our guard against supposing that such a habit could evolve along only one line. Kittlitz's sand plover kicks sand over its eggs on leaving them, and a Patagonian seed-snipe hides its eggs by scraping dry earth over them (Belcher, 1936). The practice may have arisen through the clumsiness of departing birds and been perpetuated because this chance maladroitness contributed to the protection of the eggs. Perhaps even the mot-mot's habit of plucking the barbs from its tail feathers, thus creating two oval extremities, arose through nervous agitation and was retained racially because the distinctive appearance had some biological value (Salvin, 1873).

We may conclude that nest-building arose through emotion seeking and finding an outlet in ways which were at first aimless or destructive; but even the venting of spite on the harmless grass may be formalised so that it becomes valuable in assisting the survival of the species. A primitive gull which, when excited, pulls grass and packs it around its breast is more likely to survive than a primitive raven which plucks it to throw away. Johnson (1941) discovered that northern guillemots which had idly collected pebbles while brooding thus raised their eggs slightly and saved them from destruction by flood. Selous's theory is sound if we formulate it in such terms as these: Intense emotion, not necessarily wholly or exclusively of an erotic nature, stimulates movements which in the course of time may become fixed and adapted to foster the survival of the species by furthering the construction of a nest in which the eggs and young are safer than they would otherwise be.

CHAPTER III

COURTSHIP FEEDING

Symbolic and utilitarian feeding—Courtship feeding amongst insects and spiders—Fish-presentation—Similarities between begging reactions of hungry chicks and solicitation by female birds—Evolution and function of courtship feeding.

THERE are many other species of bird besides the gannet which indulge in some form of 'billing', or 'fencing'. These endearments vary from the 'nebbing' of parakeets, the 'bill-sparring' of herons and the holding of bills by kingfishers, to the aerial 'kisses' of peregrine falcons. It is but a small step from the contact of bills to the transfer of food, so it is not surprising to find that some birds which 'bill' their prospective mates also feed them during courtship or whilst incubating. A graduated list might be compiled ranging from species which merely 'bill' lightly like the black guillemot to the rhinoceros hornbill, in which the female is entirely dependent for food during incubation on the ministrations of her mate. Whilst it is rare for feeding to take place only during the incubation period and not during courtship, commonly courtship feeding is continued in some degree during incubation, although there are a few species such as the wood duck and button quail (Butler, 1905; Seth-Smith, 1907) in which the feeding apparently ceases before incubation begins. From these facts the inference may be hazarded that courtship feeding did not arise from the habit of feeding the incubating mate but vice versa. It is significant that with very rare exceptions it is the male which feeds the female even in those species in which the cock bird shares in brooding the eggs, thus indicating that the psychological and emotional implications and effects of the ceremony are important rather than the food value of the nutriment conveyed.[1] This is borne out by the observations of Goethe (1937) on the herring gull. A female just returned from feeding will beg from her mate, although he has remained in occupation of their territory without opportunity of obtaining food. A robin, too, will importune her mate to feed her even when she is standing on a food tray in the midst of live meal-worms or has her bill full of insects for the young (Lack, 1939a, 1940a). A similar incident has been noted in the case of the reed warbler (Howard, 1907–14). A rook will present what is sometimes a make-believe meal, and symbolic feeding has also been recorded of parrots (Hampe, 1937) and the black-chinned jacamar (Skutch, 1937). The symbolical nature of courtship feeding is illustrated by the robin. When the cock bird feeds his mate he presents one insect to her; when feeding the young, to whom the actual food is important, he carries several insects (Lack, 1939a). The male Adélie penguin holds the female's bill during copulation but no food is passed (Roberts, 1940a). Darling (1938) regards the 'mock feeding' which occurs in herring gulls at the courtship stage as 'of an excitatory nature, symbolical of a later stage in the breeding cycle when the male does feed the female'.

Amongst insects there are food-presentation ceremonies both utilitarian and symbolic. Flies of the family Empidae make a bubble of silky material secreted by the male, and having embedded some prey in it present it to the female. Thus her predatory impulses are diverted from her suitor. In some instances

---

[1] Tinbergen (1931, 1932) records that in the early stages of food-presentation behaviour either male or female tern may feed the other and Lack has noticed reversed feeding amongst robins. The female button quail feeds the male, but in *Turnix* nearly all sex behaviour is reversed. A hen greenfinch has been seen to feed her mate with regurgitated food (Coward, 1923). Hudson (1913) gives a number of instances of disabled birds being fed by other species.

such gifts may be of biological value to the female, supplying her with necessary protein, but the same cannot be said of the presents bestowed by some species, for flower petals which have no food value are fixed in the bubble (Gruhl, 1924; Richards, 1927) and in other instances an empty bubble is presented. The symbolic gift functions apparently, as it does amongst birds, as a kind of sexual stimulant or psychological aphrodisiac. The male tree-cricket *Oecanthus* has a gland upon his back, and during courtship offers some of the sweet liquid which it secretes to the female (Jensen, 1909). The amorous male spider *Pisaura mirabilis* presents the female with an insect which she is still engaged in sucking when they part (Bristowe, 1929). He may suck it himself before offering it to her (Locket, 1923).

The male pigeon feeds the female before coition. Males of the yellow-billed cuckoo and Galapagos finch feed their mates while copulating (Lack, 1940*a*), and the bittern regurgitates food into the female's mouth during the act (Yeates, 1940). The female Malay coucal is fed after what Chaucer called 'a spring observance' with an insect which the male has been holding in his bill; similarly the road runner gives the female a lizard after the act. He has it ready beforehand and the female begs for it with quivering wings like a young bird (Rand, 1941). In the nuthatch (Niethammer, 1937–8) and herring gull (Goethe, 1937), coition and the presentation of food are associated.

Observation of terns shows that the gift of a small fish to the prospective bride plays a part no less important than many a bouquet or box of chocolates in human affairs. The fish is no more a mere article of food than are the chocolates; indeed, much less so, for the tern seldom gives his present gracefully and the fish is often torn in two in an unseemly tug-of-war. Here is a characteristic instance:

A tern having descended with a fish stood in the usual 'display' attitude with beak pointing downwards. She dropped by his side, 'displaying' in the usual way, except that her beak pointed upwards. Then she snatched at the fish, and in the following struggle it parted. The male walked away with his half fish and continued to 'display', clucking all the time as they do to their young. Afterwards he flew, still carrying his moiety. (Marples and Marples, 1934.)

A considerable amount of ceremonial usually accompanies the delivery of the tern's booty:

> He alights, erects his tail stiffly, half spreads his wings so that they form with his back a more or less continuous surface, something like the top of a flat-iron, erects his neck just as stiffly as his tail, points his beak heavenward, and with the fish hanging therefrom, patters round his mate with precise little steps and a lofty air of doing something more than usually meritorious. (Kirkman, 1911–13.)

Fish presentation is performed by common, Arctic, Sandwich, roseate and little terns before any eggs are laid and is thus definitely a courtship ceremony. As with the robin, in the earlier stages the offering is important in furthering the emotional bond between the pair, although authorities differ as to how far the presentation is connected with copulation (Steinbacher, 1931; Dircksen, 1932; Southern, 1938). Only later when the female is brooding does fish-presentation play a useful part in nourishing the bird (Tinbergen, 1931). In contradistinction to what obtains amongst terns it may be noted that although the female Irish dipper will perch in the nest while the male is feeding the young she makes no attempt to secure the food, but tugs-of-war may take place outside the nest when one or both birds are carrying food (Rankin and Rankin, 1940). Gulls do not act as ceremoniously as terns, but sometimes male and female seize each other's bills and the male disgorges food for the female. He commonly coughs it up for her on to the ground just as the adult birds do for their chicks. Harriers bring prey to their mates before incubation has begun, passing it to them in the air as when providing food for the young.

It is most significant that, as the quotation given above shows, food-presentation may be accompanied by the call which is used for the young. This and other observations indicate that the ceremony may be a kind of infantilism, the reappearance so far as the female is concerned of her behaviour as a chick (Holzapfel, 1939). When watching the courtship of rooks, Yeates (1934) was struck by the resemblance between the attitude and movements of the female and those of a young bird as she begged for food before copulation. He writes: 'Her whole bearing during the proceeding is one of ecstasy. But above all it is suggestive of the young being fed by the parent.' Yeates's contention that the rook plays the part of the anticipated chick is, of course, wide of the mark, for as Bierens de Haan (1929) puts it, 'The animal lives in the present, is bound by the impressions of the moment...the memory of the animal has a more dependent character than ours and is bound up with sense-experience....And no more than in the past does the animal live in the future....The animal lives in the present and in a present of rather short duration.' It is uniquely a human characteristic,

> To look before and after
> And pine for what is not.

We may not suppose that a nesting bird has any image of the chick as the consummation of its activities.[1]

In his account of the robin Lack (1939 a) says: 'The first

---

[1] There can be no sound discussion of bird psychology except on the basis of the canon of interpretation formulated by Lloyd Morgan (1930): 'We should not explain any instance of animal behaviour as the outcome of higher mental processes, if it can fairly be interpreted as the outcome of mental processes which stand lower in the order of mental development.' This 'Canon of the Minimum Antecedent' (Kirkman, 1937) is but a more precise formulation of Wundt's (1912) dictum: 'We should always have recourse to the simplest explanation possible', and this derives from the medieval rule 'Occam's Razor'.

copulation and the first "courtship-feeding" occur within a day of each other, but the two are quite separate from each other. The female invites the male to feed her with the same attitude with vigorously quivered wings and loud call as that in which the fledgling begs for food; the male comes with food in his beak and passes it to her, the female swallows it.' Many other writers emphasise the resemblance between the begging female and the actions of an importunate chick. The hen chough is fed at the nest 'drooping her wings like a young bird' (Coward, 1920) and the procedure at a red-backed shrike's nest is somewhat similar when the male feeds his incubating mate. In both these species courtship feeding occurs. In courtship the male pigeon regurgitates some secretion from his throat, probably the crop gland, and this is taken by the female in much the same way as the squab receives its food.

Not only are the actions of the female like those of a young bird but the call-note *of* the young is reproduced by the female robin and the call-note *to* the young by the male tern. This corroborates the other evidence that in courtship feeding the female bird is harking back to her youth and recapitulating the behaviour of the chick. In youth she expressed her desire for food by a certain attitude of body, quivering wings and a distinctive call; in maturity the craving which she feels is for sex satisfaction, but she and her mate approach it through the channels formed by the earlier desire. The physical manifestations of the desire for food may thus be utilised later to invite the satisfaction of other appetites.

Selous (1905a) has suggested that 'billing' is a survival of the feeding of one bird by the other, and that even kissing in the human species has this origin, and Lack (1940b) thinks that in such birds as herons it has possibly developed from the passing of nest material from the male to the female; but the origin of the practice is probably to be found in an activity more fundamental

PLATE V

page 34

RED–BACKED SHRIKES

The female is protecting the young from the sun. The male passes food
for the chicks to her. He feeds her on the nest during incubation.

page 116

HAWFINCH

The female greets her returning mate with open bill and fluttering wings.
When he gives her food, she feeds the chicks.

PLATE VI

BLACK GUILLEMOTS CHASING ONE ANOTHER

As the bird calls the red mouth is displayed.

page 40

MOUTH OF YOUNG ROOK

The yellow papillae are conspicuous against the plum-coloured background.

page 45

than either of these. The essential fact is that both 'billing' and courtship feeding are a recrudescence of an infantile mode of behaviour. Watch a young gannet persuading its parent to allow it to feed and you will notice that as the old bird dodges to avoid this importunity there is a semblance of 'fencing' which looks like the clumsy prototype of the mature birds' display. The activities of a young snake-bird in persuading the parent to allow it to plunge its head into her gullet are very similar to the 'fencing' of gannets. Here we have some of the raw material out of which courtship displays have developed. 'Billing' and courtship feeding are related and both have evolved from the motions of young birds seeking food from their parents. Gaping and reaching for food become latent and reappear as adult sexual ceremonies. We shall see later that there are other types of courtship behaviour which can be interpreted as a recrudescence in more or less conventionalised form of the reactions of the chick desirous of food.

There can be no doubt that the function of courtship feeding is to stimulate and maintain the emotional relationship between the pair of birds. Whenever it occurs it is accompanied by emotional excitement (Tinbergen, 1936a). It is significant that, for example, courtship feeding occurs in the bob-white, a bird which maintains a pair bond, but that with a few exceptions such as button quails it does not occur among the many Gallinaceous birds which do not remain paired. Sexual faithfulness has biological value, as young broods guarded and maintained by both parents have, in many cases, a greater chance of survival than when they are tended by one parent only. And if courtship feeding has such importance in maintaining the pair bond we cannot doubt its efficacy in furthering the synchronisation of the bodily rhythm of the two organisms on which the whole process of generation depends.

CHAPTER IV

CEREMONIAL GAPING

Ceremonial gaping of British sea-birds, birds of paradise and other species—
Examples amongst snakes, lizards and fish—Gaping as intimidation—The
biological value of the adornments in nestlings' mouths—Examples of
precocious behaviour—Adornments connected with feeding associated in
later life with sex—The importance of surprise in stimulatory display and
the necessity of effective display for successful breeding.

THE reader will recollect that amongst the remarkable
gestures of the gannet is the opening wide of the
mandibles to reveal the black interior. We may well
devote some attention to an enquiry into the significance of this
performance. It is characteristic of a great variety of birds.

From November onwards until June red-breasted mergansers
swim in small 'displaying parties', and long ere the drakes are
in full breeding plumage they posture vigorously (Armstrong,
1940). With the warmer spring weather their fussiness increases.
There is continual turmoil and excitement in the flocks—ducks
flying up and away with drakes in pursuit, a constant coming
and going, fluttering and splashing. Now they joust with each
other, rushing at their opponents with open bill and threatening
mien, now paddling around displaying their gorgeous apparel
to best advantage they turn to chase and woo the coy females,

> Endeav'ring by a thousand tricks to catch
> The cunning, conscious, half-averted glance
> Of their regardless charmer.
>
> J. Thomson, *The Seasons*

The drake swims towards the apparently unheeding duck
with carmine bill raised at an angle of seventy-five degrees and
neck stiffly held in line like a menacing snake; then, as he reaches

her he executes a sudden bow, emitting a soft scraping or purring note. He raises his wings, extending the black-barred white secondaries, and stretching his neck, displays the neat white collar. Then, rising in the water to show off his breast-band, he dips again. And now comes the strangest part of the performance. With magnificent crest fully spread the drake turns his head, twisted at an angle of some sixty degrees to the water, towards the female, and opens his beak to its fullest extent, revealing the brilliant red interior (Millais, 1913).

This display may vary in a number of ways. A female, impatient of the attentions of the drakes, may dive and swim away under water. When she reappears one or more of the drakes rush towards her, sometimes with neck retracted, sometimes with it fully outstretched. On reaching the duck the head and neck are jerked upwards almost simultaneously with the tilting of the rear portion of the body, and on returning to their normal position the crest is raised and the beak opened. The extent of its opening varies a good deal. The performance is not to be thought of as being leisurely and dignified; it happens very suddenly and may take no more than two seconds to execute. The sequel will make it clear how important it is that, if possible, the duck should be startled by the drake's efforts.

On one occasion seven birds, six of which were drakes, flew over my head and alighted in a sheltered shallow creek in an Irish lough. With crests fully erected they darted excitedly at each other and pursued the duck. Every now and then one would jerk up head and neck at an angle of fifty degrees, and with the neck, so far as I could ascertain, slightly concave to the water during the sudden movement. This action continued to be performed occasionally even after the duck, pursued by a drake, had left the party, and seemed to be partly an expression of rivalry but perhaps more the result of nervous tension. At this juncture, one drake, as he swam, plunged his arched neck below the surface six times in close and regular succession. Bills were occasionally opened and the interior displayed.

The mergansers' scufflings are very energetic and cause much splashing, but I have never noticed any fighting with the formidable bills. In love they are used to impress, and in war to intimidate (Townsend, 1911; Colman and Boase, 1925; Phillips, 1922–6; Richmond, 1939).

Ceremonial gaping is commoner amongst birds which frequent the water than amongst land birds, at least so far as the British Isles are concerned. No bird exemplifies the curious practice better than the shag, so let us imagine ourselves concealed near a group of these birds as they stand in various grotesque attitudes on a rock ledge. One of them is obviously excited. He approaches a female with just such a trace of alacrity in his step as to suggest that he might begin to dance if he were a little less clumsy. He gives an unexpected and unwonted hop, stretches up his sinuous neck, throws back his head in line with it and opens wide his bill for a second or two; then closes it, and opens and closes it several times in succession. He now sinks forward on the ledge and tilts his stumpy expanded tail forward so that his bill, thrust over his back, is amongst the feathers there; and thus he remains in a kind of ecstatic posture. The amorous bird may repeat the performance several times. When his head is over his back, the throat is seen to vibrate as if some sound were being uttered, but nothing can be heard. The yawn reveals the vivid coloration of the buccal region; not red as in the red-breasted merganser but bright gamboge yellow like the naked outer skin at the base of the bird's mandibles.

But it is not always the male bird which acts in this way. Indeed, Kortlandt (1938) has shown that in the display of the southern cormorant the rôles of the sexes are interchangeable, and probably this also applies to the display of the shag which is thus described by Selous (1927):

May 16th. A female shag flies in to the rock, accompanied by a male. He follows her to the nesting site, and here now, first squatting herself down as a necessary preliminary, she solicits him violently in full measure. He presses against her, caresses her with his bill, and seems as though about to respond more fully, but fails to do so. Then, as has been the case before (I have to chop), another male flies in, and she now, in the most unmistakable manner, solicits *him*. There is no doubt whatever as to the addressee, for she keeps shooting her head forward towards the newcomer and away from the other (who, though still juxtaposed and affectionate, is *behind her head* by some six inches), gaping her bill each time that she does so, which is as if to flash a bull's-eye lantern at him—her buccal lantern, one may well call it. He who has just before been her dear does not at first seem quite to understand the revulsion, makes some mild advances, divines that there is something wrong, desists and looks foolish. As for the other—the *bienvenu*, now being furiously wooed—he is all hesitancy and bashful faint-heartedness, but in the most interesting manner is lured or, as it were, magnetically drawn to her by the continued flashings and quiverings—the latter now also a very marked feature—of his inamorata. He comes a little way up the rock, wavers, looks shy and stops, then comes a little farther up it, stops again, again comes on, each time in clear response to the solicitations, which get more and more energetic and are especially accentuated on every fresh start forward that he makes. He is now fairly near and I am expecting fuller developments, when the partner of the female, near by, on the nest—possibly mistaking the direction of his advance—comes *hopping* heavily down the rock, and, with a run at him, puts him to flight.

The whole passage is worthy of particular attention, illustrating as it does exactly how sexual attraction is exhibited and the way in which it operates. It is as certain as anything can be in these matters that the revelation of the yellow mouth has a stimulating or seductive effect on the bird which sees it.

As kittiwakes stand on their ledges male and female crane their necks towards one another and scream with bills widely opened; and lo! the interior is revealed as of a lurid orange-red. Even when the shriek has died away the mandibles remain apart and the craning motion continues. Obviously the value of the opening of the bill for the birds is related to the display of colour as well as the utterance of the call; so also with fulmars which constantly indulge in bouts of head-shaking with distended mandibles. The cavity is tinted bluish mauve (Selous, 1905*b*). A hoarse *eck, eck, eck* sometimes accompanies the display, and occasionally as the female holds her bill open the male may transfer some oily matter to it (Coward, 1926; Boase, 1924). Razorbills, puffins and black guillemots exhibit their vivid interior mandibles as they posture. The first two species show yellow and orange colouring respectively, the latter red. By analogy we may surmise that the great auk had some such gaping ceremony which no ornithologist will ever behold.

There are other birds of the sea and fresh water which gape during courtship. Many of the ducks do so, though no others, so far as I am aware, quite so extensively as the merganser and goosander. In the hooded merganser's display the crest is prominently exhibited and the bill is opened to emit a double note rather than to show mouth-coloration (Bagg and Eliot, 1933). It is noteworthy that the saw-bills illustrate the correlation of adornment and display-movements very clearly. Red-breasted mergansers have vivid mouths and gape widely, the smew rears up with open bill and shakes his head from side to side (Hollom, 1937), but the mouth display is subordinate to other movements showing the crest and strikingly contrasted plumage; the hooded merganser makes great play with his elaborate crest and gaping has become unimportant. The sheldrake opens his bill wide during nuptial preliminaries, but most commonly as a means of intimidation. Sandwich terns will face each other on the shore and call together with wide-open beaks held high and then parade in circles, first in one direction, then in the other (Marples and Marples, 1934). Great crested grebes, too, will front one another, opening their beaks like a pair of scissors at fullest extent (Selous, 1933), and ypecaha rails indulge in communal concerts and dances during which they rush about with wide-open bills pointing vertically upwards (Hudson,

1892). Lovesick storks also demonstrate with expanded mandibles (Hingston, 1933).

Wrynecks have a mutual display, not confined to courtship, during which the male and female perch opposite one another, throw their heads back and shake them up and down with the mandibles distended to show their pink mouths. Sometimes the head is allowed to hang down limply (Thompson, 1938). This species is one of the birds which 'feigns death' when taken in the hand—lying limply with closed eyes. Later, when we consider the possibility of a connexion between courtship display and 'injury-feigning' this resemblance should be borne in mind.

A number of other British birds have more or less vaguely defined 'yawning' displays—the marsh warbler, blackcap, yellow bunting, blackbird and thrush. The cuckoo, great and blue tits, puffin (Perry, 1940), and many other species threaten opponents with open mouths. The raven displays its maroon-coloured gape in menace as well as courtship. Gaping, indeed, in this species and probably in others, is one of a number of actions which assist sex recognition. The response of the female is different from that of the male. The one is attracted, the other roused to anger or terrified into retreat.

Gaping is characteristic of several of the birds of paradise. Their social displays, or *sacaleli* were described by Wallace (1869), but for accurate accounts of the posturings of individual birds we are dependent on notes made of specimens in captivity, such as the description of a king bird of paradise by Sir William Ingram (1907):

He always commences his display by giving forth several notes and squeaks, sometimes resembling the call of a quail, sometimes the whine of a pet dog. Next he spreads out his wings, occasionally quite hiding his head; at times, stretched upright, he flaps them, as if he intended to take flight, and then, with a sudden movement, gives himself a half turn, so that he faces the spectators, puffing out his silky-white lower feathers; now he bursts into his beautiful melodious warbling song, so enchanting to hear but so difficult to describe. Some weeks ago I was crossing a meadow and heard the song of a skylark high up in the heavens, and I exclaimed at once: 'That is the love-chant of my King-bird'. He sings a low bubbling note, displaying all the while his beautiful fan-like side-plumes, which he opens and closes in time with the variations of his song. These fan-plumes can only be expanded when his wings are closed, and during this part of the display he closes his wings and spreads out his short tail, pressing it close over his back, so as to throw the long tail-wires over his head, while he gently swings his body from side to side. The spiral tips of the wires look like small balls of burnished green metal, and the swaying movement gives them the effect of being slowly tossed from one side to the other, so that I have named this part of the display the 'Juggling'. The swaying of the body seems to keep time with the song, and at intervals, with a swallowing movement of his throat, the bird raises and lowers his head. Then comes the finale, which lasts only a few seconds. He suddenly turns right round and shows his back, the white fluffy feathers under the tail bristling in his excitement; he bends down on the perch in the attitude of a fighting cock, his widely opened bill showing distinctly the extraordinary light apple-green colour of the inside of the mouth, and sings the same gurgling notes without once closing his bill, and with a slow dying-away movement of his tail and body. A single drawn-out note is then uttered, the tail and wires are lowered, and the dance and song are over.

During his display the superb bird of paradise erects a ruff around his head, clicks his wing-feathers and opens a bright green mouth (Stonor, 1940). In the case of Hunstein's magnificent bird of paradise the male exhibits a mouth of a delicate apple-green shade (Seth-Smith, 1923). Rand (1940 a) watched a wild magnificent bird of paradise displaying his yellowish-green mouth after coition had taken place. The twelve-wired and long-tailed birds of paradise perform gaping antics, the former showing bright yellow, the latter green (Crandall, 1932, 1937). Wilson's bird of paradise when in display utters a churring sound disclosing at the same time light green mouthparts (Winterbottom, 1928). Another member of the Paradiseidae, the rifle bird, includes mouth opening in his spectacular display. During his ecstasy he throws his head to one side so that it rests on the wing and sometimes opens his bill revealing his shining white gullet, thus adding another wonder to a galaxy of splendour (Selous, 1927).

No doubt there are many species with gaping ceremonies which have not been mentioned, but this short review of a strange custom has shown that there is no colour which may not lie concealed in a bird's throat; black and white, red, orange and yellow, mauve and bright green are all represented.

The exhibition of a lurid mouth is by no means confined to birds. It is not unusual amongst reptiles, especially lizards, but always as a threat and not in courtship as has been suggested by Pycraft (1914). The researches of Noble and Bradley (1933) have shown that the vivid colours of lizards are used for purposes of intimidation and that courtship in the proper sense hardly exists amongst these animals. When excited the frilled lizard (*Chlamydosaurus kingi*) stiffens its frill and opens its mouth, which at such times becomes flushed with vivid red. A frightened moustached lizard (*Phrynocephalus mystaceus*) stands on its hind legs with jaws as widely apart as possible, displaying a pink mouth and presenting a most horrific spectacle. An arboreal lizard in Portuguese East Africa, *Agama atricollis*, menaces enemies with a brilliant orange-yellow mouth (Cott, 1940). Frightened chameleons bluff their enemies by gaping at them. Crossland (1911) writes thus of one which was sometimes chased by a fox terrier in a house in Zanzibar:

> The chameleon invariably tried to run away when attacked, but those who know the species can imagine the ludicrous ineffectiveness of a chameleon's flight. In a few seconds the impossibility of escape seemed to reach the animal's brain, when it at once turned round, opened its great pink mouth in the face of the advancing foe, at the same time rapidly changing colour, becoming almost black. This ruse succeeded every time, the dog turning off at once.

The opening of the mouth in menace has been recorded of snakes such as the pink-mouthed green tree snake of India and

Ceylon *Dryophis mycterizans* (Wall, 1921) and the American hog-nosed snakes (*Heterodon*), the pike-headed snake (*Oxybelis acuminatus*) and the Fer de Lance (*Bothrops atrox*) (Cott, 1940). Some fishes, such as *Tilapia heudeloti*, *T. dolloi*, and *T. natalensis*, try to intimidate each other, mouths agape.

These menacing demonstrations are analogous to the antics of some birds already mentioned, such as the sheldrake, merganser, raven and cuckoo. The Australian frogmouth endeavours to frighten intruders by erecting the feathers on its crown, drooping its wings and opening its huge beak so that the interior of the mouth is displayed (Chisholm, 1934).[1] When a nightjar is defending its young 'it transforms itself into an almost repulsive object, as, with head thrown back and the huge red gape wide opened, it slithers along the ground towards the intruder, beating the earth with its wings and feigning mortal injury' (Turner, 1914). Young nightjars also distend their mouths in menace when disturbed. An incubating dusky poorwill opens a cavernous pink mouth and hisses (Aldrich, 1935). Possibly the intimidation originally lay, not in any association in the other bird's mind between the colour of the mouth and danger, but in the threat of being seized by the adversary's bill, for many birds and lizards when fighting do clasp each other's bills or jaws, and we may argue that the already existing mouth colour acquired warning connotation, or that by the process of selection the colour was developed or accentuated because of its value in frightening away or cowing an enemy.

The spectacular effects in the mouths of young birds are not less remarkable than those of their elders. In the nestling

[1] Writers refer to the vivid hue of the mouth as if it played a part in accentuating the menacing aspect. Whether it intensifies in colour seasonally or with age I do not know, but in summer the specimen in the London Zoological Gardens has beak and mouth of a uniform dull horn colour. The mere sight of the bird alarms other species; Pekin robins chatter in terror when one is shown to them (Dewar, 1928).

bearded tit's mouth there are four rows of little pearly white conical projections of unequal size, two rows on either side of the palatal cleft. These tooth-like bodies are set against a background of black, which in turn is surrounded by carnelian red. The whole is fringed with yellow, provided by the gape-wattles. The tongue is black with two white spurs at its base and a white tip. In the case of chicks of the South American ash-coloured cuckoo, the mouth cavity is outlined 'in a startling manner' with white tubercles (Wetmore, 1926). The young hawfinch's mouth is decorated with a striking red and blue pattern. The skylark chick has three black spots arranged in a triangle on the tongue and another black mark near the tip of the lower mandible. A rook nestling's mouth changes from bright cinnamon to a dark plum colour as the bird develops, a kittiwake's from pale pink to rich red or orange-red, a fulmar's from pink to mauve, a shag's from pink to gamboge yellow. A young cuckoo's gape alters in nine days from yellow to red (Armstrong, 1929). Exotic birds exhibit various kinds of brilliant colouring and odd adornment about their mouths. The Gouldian weaverfinch has fine black spots on the roof of its mouth and a tongue crowned with a black stripe. At the angle of the gape are globular excrescences of bare skin coloured brilliant emerald-green and blue, and these are supplemented with outgrowths of brilliant opalescent green and blue skin (Pycraft, 1934). The red-tailed weaverfinch of Samoa has bright red gape-flanges. Chisholm (1934) writes of the finding of two young Australian scaly-breasted lorikeets with luminous mouths, but this record should be accepted with reserve.

Leaving aside for the moment the question of the meaning of the vivid hues in the mouths of adult birds let us ask, What is the significance of the bright markings and strange decorations revealed in the mouths of nestlings? There can be little doubt that the function of these is to indicate to the parent bird where to place their chicks' food. What could be more convenient than that when the busy parents arrive with their dainties they should find the mouth cavity vividly painted or outlined so that they can dispose of their gleanings quickly and accurately and then rush off for more with the minimum delay? The gape or gape-wattles mark out the circumference of the lively receptacle, the bright colour and spots or stripes inside provide an additional series of clues. Thus, there is little danger of the most hurried or careless bird dropping the food wastefully by mistake into the bottom of the nest. This explanation is supported by the fact that on the whole it is the young of birds born in relatively dark places and remaining for a considerable time in the nest which have decorated mouths. Nidifugous chicks do not have remarkable mouth adornments. The theory works both ways, for the young gannet, cormorant or fulmar is able the more readily to find the gullet of its parent into which to plunge its bill by reason of its coloration; and this is such as to be peculiarly conspicuous by reason of its difference in hue from the general tone of the plumage. What could contrast more vividly with the cormorant's dark garb than its gamboge yellow mouth; and how black is the gannet's gape compared with the bird's snowy feathering! It is interesting to note that the herring gull, a bird which disgorges food for the young, has a red marking at the base of the bill which apparently stimulates the chicks to peck there in search of food (Goethe, 1937).

The tongue spots occur on places specially sensitive to the touch and serve the double purpose of guiding the parent in placing the food and ensuring the closing of the mouth when the food is administered. I believe, however, that these mouth decorations have an even more important role to play than that of 'indicators'. They are also 'stimulators' and have the function of releasing the impulse to feed the young. Just as the diffident, shy shag was drawn to the amorous female by her bright gapings, so the parent bearded tit, warbler or finch is induced by the buccal display of its offspring to disgorge its catch. Proof of this is furnished in a negative way by the observed instances when one of a brood, unable to compete with its brethren for food, becomes too weak to beg for it and is allowed to die. Positive proof is given, as Lorenz (1935) points out, by the mouth coloration of young cuckoos. The mouths of the birds which evict their nest-mates and thus have the nest and the foster-parents' attentions entirely to themselves do not resemble the mouths of their fosterers' chicks, but in those species in which the cuckoo grows up amongst the fosterers' progeny there is similarity of mouth coloration. It is significant, however, that during the European cuckoo's first few days its mouth most nearly resembles those of its fosterers' young. It is well known that the begging actions of the cuckoo have an almost compulsive effect in inducing other birds to feed it. A striking instance of this occurred when a South American black tanager squeezed its way into a cage containing one of these young birds in order to feed it (Finn, 1919). Specific markings about a bird's body have the power to initiate specific responses; the colouring of cuckoos' mouths illustrates this very clearly. Unquestionably the mouth markings of nestlings as well as the begging movements have an important function in inducing the parents to feed them.

Both aspects of the theory find support in the remarkable difference of coloration in the mouths of the male and female rhinoceros hornbill—black in one case, pale fleshy in the other (Darwin, 1871). The female is walled into the nesting cavity and is fed by her mate. As the cock bird does not need a mouth indicator, the colour in his case is less pronounced, but so far as feeding is concerned, during the period of incubation the hen is as a young bird and she can be more readily fed by reason of the pink hue of her capacious mouth. We may say too, following the theory just stated, that the coloration tends to induce him to feed her.

Thus a foundation is laid for mouth coloration to have a special emotional meaning associated with the contentment resultant on the giving or receiving of food. It is reasonable to regard the satisfaction originally aroused in the species by the feeding of the chick as responsible for the many pre-nuptial ceremonies in which the male feeds the female and the pair 'neb' together. The practice has probably evolved to the advantage of some species, such as the rook, in so far as the male by making himself responsible for the feeding of his mate enables her to guard and tend the eggs and young much more diligently than would be possible if she had to forage for her own subsistence. The close connexion between the two fundamental desires, food and sex, is vividly shown in an excerpt from Kirkman's (1937) study of the black-headed gull:

> The only marked preliminary to the sex act is the 'Kwurps', and less often, the 'Kwarrs' of the male gull. As the usual occasion is the feeding of the hen, and as the hunger note is also 'Kwurp', it is possible to see the pair standing or walking together, both 'Kwurping', one moved by hunger, the other by sex. Usually, but by no means always, the hen is fed first; the sex act follows.

Nature likes to use material ready to her hand, and when the time comes for overtures towards the satisfaction of sexual desire she uses sound, colour and movement which have already an emotional content through their relationship to another appetite.

We have already noticed numerous illustrations of this; many birds in their courtship rite offer food; some, like ravens, hold each others' bills in a lengthy 'kiss'; others, such as doves, bow in a way which reminds one of their actions in regurgitating food for their young; yet others, like gannets and albatrosses, fence with their bills in a manner suggestive of the chicks seeking food from the parent; many cower with fluttering wings or hold forth one quivering wing as when being fed. There is much evidence that the reactions of the young are part of the raw material out of which elaborate courtship displays are evolved, or, at least, that they contribute important elements. For example, the nestling fulmar petrel waves its open bill from side to side suggesting the 'yawning' gesture of later life. In commenting on this, Selous (1905b) adumbrates the theory developed in this book. He says: 'May not the habit have originated in the hunger of the chick, and have been worked in sexually at a later age, when the reproductive system had become active? Strong emotion, one may suppose, would require an outward manifestation in the shape of movements of some sort, and it would be such as were already known, that, by first coming to hand, would be likely to be first employed.' It must be pointed out, however, that not a few precocious activities such as were mentioned earlier—collecting nest-material, brooding and displaying— constitute, not behaviour-patterns in process of evolution but manifestations of incomplete behaviour-patterns due probably to the untimely activity of the gonads. Bullough points out in a paper on the starling (in the press) that it is only after the young have left the nest some considerable time that the gonads regress, and if this be true of birds in general showing precocious behaviour it would provide a physiological explanation of it. The reciprocal influence of psychological and physiological states is so complex that it is often difficult to decide which is primary either in evolutionary development or functional significance.

If the preceding argument be sound, we are justified in regarding the mouthparts of birds as possessing an emotional connotation from their earliest days. The satisfaction of the hunger of the young cormorant, shag or gannet is intimately associated with the coloured gape of the parent. Also, as we have seen, amongst a great variety of species the male feeds the female at a certain stage in the sexual cycle. Thus the beak and mouth become saturated with emotional significance. Moreover, as birds commonly open their mandibles in menace when intimidating a rival, here is yet another means whereby gaping may acquire emotional content.

When we consider these facts it should not seem odd that the mouth and adjacent regions should play such a large role in the ceremonial of birds. They are the parts of the body which earliest and most definitely are associated with the satisfaction of desire. The Freudians, indeed, have been at great pains to point out how true this is of human beings. In later life they are employed to express and induce emotions of various kinds. From this standpoint we are now able to understand the gaping ceremonies which have been described.

The effect produced is a heightening of emotional tone. No doubt of this can be entertained by the careful reader of Selous's (1927) narrative of the solicitations of the male shag by the female:

I have seen the male shag, whose whole aspect speaks of timidity and nervous embarrassment, drawn up the steep slope of the rock, as it were, by an invisible thread, which is tightened with every increase of the often quite frantic demonstrations of the enchantress—this Circe—who holds it, and relaxed as these sink. Thus, by a few steps at a time, he comes gradually nearer and nothing can be more noticeable than the joy of the siren and the strengthening of her efforts at each renewal of these advances, which, at first faint and timid, become bolder by degrees, and love's victory seems well-nigh won. But now there comes a pause, lure and response fail together, he stands still, glances backward and seems to repent. He would go if she would let him, but whilst he yet hesitates she awakes as from a sort of swooning ecstasy. The charm, and with it the yielding, is renewed, and so again and still again—each time more wildly, more deeply—till there is no longer the wish to escape.

Selous, as we have seen, aptly calls the exhibition of the interior of the mouth 'flashing the buccal lantern', and by doing so he draws attention to the dramatic nature and suddenness of the gesture. It is thus also with the red-breasted merganser and other species which have been mentioned. Suddenly the bird's mate, or prospective mate, finds himself, or herself, looking into a chalice of flaming splendour. Even from the standpoint of a human observer the sight is startling enough; how much greater must be the effect on an eye only a few inches away and adapted to respond to this display!

One can hardly over-estimate the importance of the factor of abruptness in provoking a bird's reactions to change as rapidly and completely as when a switch is pressed. Some stimulus, visual or tactile, occurs—no doubt correlated with specific psychological susceptibility—and behaviour alters *toto coelo*. This is not merely a matter of startled reaction but the setting into motion of a new behaviour complex. A noddy tern, for example, flies away from the nest when disturbed if there are no eggs there; but place eggs in the nest and immediately on the bird's perceiving the alteration she conducts herself in quite a different way, becoming anxious to brood and bold in defence of these alien eggs (Watson, 1908). A male black-crowned night heron will want to brood immediately he sees eggs in the nest (Noble, Wurm and Schmidt, 1938). We may believe that courtship display stimulates or tends to stimulate the appropriate pattern of behaviour in a somewhat similar way.

An experiment conducted by Eliot Howard (1929) in connexion with a non-sexual response shows how conduct may alter according as the element of surprise varies. The reed bunting has two different modes of behaviour when disturbed at the nest. It may perch on the rushes with spread tail and anxious call, or it may drop to the ground and run about with wings folded, feathers raised and tail spread—a reaction related to the 'broken-wing trick', which some ground-nesting birds and others which nest near the ground adopt when their young are threatened. The first mode of action is usual when a stoat or weasel approaches the young. Howard lay hidden and threw a ferret towards the chosen nest, making ten experiments with five different birds. Only on one occasion—when the ferret fell beside the nest—did either male or female run about the ground. But the experimenter discovered that the 'ground reaction' occurred if he showed himself suddenly close to the nest. 'It was the suddenness, even more than the strangeness of the stimulation that elicited response. So, likewise, the wing suddenly extended or vibrated—the rush of the guinea fowl with out-spread wings; the skylark's wing shot suddenly over his mate; the pied wagtail's wing suddenly raised and expanded as he runs round his mate; or any of the numerous and diverse forms of posturing—whether by male or female—increases the intensity of stimulation and establishes a waning reaction or one of low intensity.' Amongst these 'numerous and diverse forms of posturing' we may include ceremonial yawning.

Consider the actions of the amorous peacock of our parks and gardens as he prepares to woo the peahen. He does not swagger boldly up to her with all his splendours displayed. Having approached within what he judges a proper distance of the lady —not too close—he slowly erects his marvellous plumes. Then he turns his back, presenting nothing but his drab wings and tail and the great brown disk of the raised train to the eyes of his chosen one. Now he walks backwards towards her, and when he considers himself close enough, whirls round, vibrating the long quills of the train feathers so that a half-rustling, half-rattling sound is heard; and to add the supreme surprising dramatic touch to the performance he utters a loud penetrating scream. Then he stands expectant before her. Everything is designed to make the effect as sudden as it is spectacular.

Great crested grebes will front each other in the course of connubial antics and suddenly rise in the water, revealing their

shining white bellies, visible hundreds of yards away. The great
bustard inflates his air-sacs, averts wing and tail feathers, and
throws his plumes into a billowy mass; when Franklin's grouse
goes into display 'the whole bird is transformed with the
suddenness of a conjuring trick' (McCabe, 1932); the frigate
bird swells the great scarlet bladder beneath his throat, droops
his wings, calls *wow, wow, wow, wow*, and clatters his robber's
beak; the Costa Rican bell bird inflates three enormously
elongated spiked wattles which spring from the base of his beak.
The strange and the sudden are employed to impress and stimulate the female because dramatic, vivid displays have a biological
value. And if we ask where a bird may most conveniently keep
concealed a patch of colour available for startling exhibition the
obvious answer is—in its mouth.

The importance of all these devices, including gaping, lies
in their effectiveness in heightening and preserving emotional
tone, and consequently securing synchronisation of the sexual
rhythm. Birds have no prevision either of the sequence of the
seasons or the organic changes in their own bodies, so the time
at which a bird migrates and breeds is determined on the one
hand by these bodily developments, and, on the other, by the
influence of light and other factors in the environment. It is
through the correlation of the factors involved that a bird
eventually breeds at the optimum season—that is, when climatic
conditions are favourable and a sufficient food supply is available. Any failure in the mechanism of adjustment may spell
disaster, either through the bird's not being in suitable sexual
condition at the appropriate time or the young brood being
exposed to unsuitable conditions.

Watch a flock of lapwings or chaffinches in spring and it will be
noticed that when a warm spell comes they tend to separate into
pairs, but if the cold weather returns they are once more seen
in flocks. The relationship between bird and environment is in

very delicate equilibrium. Sequence, too, is important, for many reactions are not reversible. Bird behaviour is a pattern of reactions such that its disorganisation at one point may destroy the integrity of the whole. Into this very intricate complex the activities and reactions of male with female have to be co-ordinated at the right moments. 'Never the place and the time and the loved one all together' spells extinction. Nature solves the supreme problem of synchronisation by a host of cunning devices. In many cases the cock bird migrates first and secures a breeding territory ready for his prospective mate when she arrives, though a considerable number of marine species appear at their breeding grounds already paired. By displaying he is able to stimulate her and accelerate her sexual rhythm so that co-ordination is achieved. The supreme desideratum is to touch the spark of passion into flame, to arouse the pitch of emotion that union may be consummated. To that end, as we have seen, the most bizarre adornments have been evolved, the strangest performances elaborated.

True, the male's antics often seem to be without effect on the apparently unheeding desired one; but careful scrutiny usually reveals that the coy creature is not so indifferent as she appears, and we should not suppose that they are futile. It may well be that each time the pantomime is performed it has some effectiveness in raising the emotional tempo, thus preparing the way for future success.

This explanation needs modification in order to apply to the birds which continue duets of 'mouth opening' and other dual rites and mutual displays after they have paired and incubation has begun. We shall have more to say about these performances later on, but this much may be emphasised here—that although they cannot in their post-nuptial manifestations have a value in initiating the union, they play a part in preserving it. What naturalist observing the dual ceremonies of whooper swans could doubt the function of these domesticities in maintaining a partnership which is happy in every sense; especially so biologically, by securing the care and guardianship of the cygnets until they are old enough to fend for themselves?

We may conclude that the principal function of gaping displays is either to initiate, augment or maintain mutual feeling. Nature uses the simple, primitive reactions, and not only formalises them but gives them special, conventional and sometimes arbitrary meaning. She loves complexity, and experiments to discover into what strange and intricate designs she may weave her simplest patterns; she does not like to discard any organ or habit until she has played a prolonged game with it, trying to discover how it may be more amply employed— even to serve ends quite other than those for which it was designed. Give her time and there is no organ so petty nor action so trivial that it may not be made part of a lovely design. She is not content to let beaks be for seizing and mouths for eating; they can be used artistically in the reproductive drama. Variety is the spice of Nature's life; and if we consider the many beautiful bypaths of form, colour, sound and activity into which she has strayed, we have reason to conclude that, like Wisdom, she is justified of all her children.

CHAPTER V

THE SIGNIFICANCE OF RELATED
CEREMONIES

The co-ordinated flight of gannets, pelicans and cormorants as a clue to their
relationship—Evidence from behaviour of the reptilian origin of birds—
Analogies between bird display and human etiquette—Similarities between
the displays of albatross, gannet, gull, wood pigeon and other species—
The phylogenetic constancy of instinctive behaviour—Behaviour-clues of
taxonomic value amongst doves and ducks—Analogies in human culture-
patterns and ceremonial.

THE reader will probably have noticed that amongst related species of birds there tends to be a general similarity
of ceremonial. Until quite recently display behaviour
was disregarded by systematists; indeed, there are few to-day
who realise how much it has to tell us about the affinities of
living creatures. For this state of affairs they should not be
blamed; the human mind deals most easily with what is con-
crete, as Dr Johnson showed when he thought to refute
Berkeley's philosophy by kicking a stone; and structure is more
readily analysable than behaviour. But now it is clear that the
interests of science will be best served by co-operation between
the taxonomist and the out-of-doors observer.

Those who have watched gannets flying to their fishing
grounds will have been charmed with their beautiful custom of
proceeding in follow-my-leader formation. As many as a dozen
or more birds regularly spaced in single file flap low over the
surface of the sea, but ever and anon they all cease plying their
wings and glide for a few seconds. So they go, alternately
flapping and gliding, an orderly unit as if under command.
Surveying these files of birds from a high cliff one summer day
I recalled that elsewhere I had seen birds acting in a similar way,

and there flashed into my mind memory-pictures of pelicans flying over an Indo-Chinese lake and a blue bay in Trinidad. They, too, intermitted their flapping every now and then to plane over the water. Thus an idiosyncrasy of flight brought home to me the close relationship between gannets and pelicans. Cormorants, classified with gannets and pelicans in the Pelecaniformes, also sometimes glide as they fly though less regularly; but in company with pelicans they adopt the same rhythm (Beebe, 1938). Association with these birds brings into evidence a mannerism common to the three Families but for the most part latent in the cormorants.

That peculiarities of movement may have significance for systematics over a range much wider than that between Families in the same Order has already been emphasised by Heinroth (1930). He draws attention to the fact that many birds lower their wings in order to scratch their heads and points out that this procedure can only be understood when we realise that the wing is evolved from the reptilian arm. Functional efficiency would be served by scratching the head while the wing remained depressed in the usual resting position on the back, but we are here dealing with a vestigial mannerism which must be interpreted historically. It is highly probable that the prevalence of gaping as a courtship antic amongst birds is another behavioural relic reminding us of their reptilian ancestry.

Courtship display may be regarded as comparable in some respects to certain forms of etiquette which have as their function the conveyance of meaning and the establishment of a particular relationship. The greeting ceremony of an African native who spits on the stranger is apt to be misunderstood by the European accustomed to extend his right hand, just as the mandrill, exposing his canines in friendly overtures to a baboon, whose greeting ceremony consists of smacking the lips, is received roughly (Bierens de Haan, 1929). Greeting etiquette can and does vary immensely because it is not functionally necessitated; therefore the anthropologist can use it as a significant factor indicating cultural contacts between peoples in the past. On a lower level, bird ceremonial, being phylogenetically constant and conventionalised, indicates past relationships and present affinities.

A comparison of the gannet's posturings with those of some other birds shows how similarity of display-pattern indicates affinity between species and genera. To avoid the suspicion of even unconsciously exaggerating the likenesses between these ceremonies it will be best to quote descriptions of them verbatim.

Crossing the Pacific Ocean from Japan to Honolulu one sees a large, tireless, long-winged bird swerving around the ship's stern. This albatross, which nests on the island of Laysan, indulges in a display remarkably akin to the antics of the gannet.

First they stand face to face, then they begin nodding and bowing vigorously, then rub their bills together with a whistling cry. After this they begin shaking their heads and snapping their bills with marvellous rapidity, occasionally lifting one wing, straightening themselves out and occasionally blowing out their breasts; then they put their bills under the wing or toss them in the air with a groaning scream, and walk round each other often for fifteen minutes at a time. (Rothschild, 1893–1900.)

This account may be compared with Beebe's (1926) description of the courtship of the Galapagos albatross:

My bird walked up to its mate, which, in its turn, rose and faced the new arrival. They stood with their breasts a foot apart. My albatross suddenly shot its head and neck straight up, the bill pointing skyward, uttering at the same time a deep grunting moan. Its partner followed suit, then alternately, each bird bowed deeply and quickly three times. Without an instant's delay they next crossed bills and with quick vibrating movements of the head they fenced—there is absolutely no other word for it—with closed mandibles. Without warning, my bird ceased, and again shot his head high up into the air. Its mate instantly turned her head and neck sideways and held them motionless and concealed from my point of view, close to the left wing and side. Then another bow and a second bout.

Later, one was seen to hold its mandibles wide open, 'the other's beak even entering its mouth once or twice'. Beebe noticed that this ceremonial continued even after incubation had started and found that a bird would go through the performance with several others in turn and even with human beings. He saw an albatross perform with three separate individuals in the course of a long and difficult journey through the colony.

Pinchot (1931) shows that gaping is part of the Galapagos albatross ritual and reproduces a photograph depicting two of the birds fronting each other at a distance of about two feet with mandibles wide apart. According to his account the 'dance' consists of three figures, bowing, gaping and fencing, and it is repeated in sequence three times or more. The beaks are not merely opened and closed, but are held open for a moment. All the figures of this display are represented in the display of the gannet. Furthermore, as we have seen, mouth opening occurs in the fulmar which is placed by taxonomists close to the albatrosses and the pelicans, and the shag, a bird closely related to the boobies and possessing, in common with some herons and storks, the extraordinary backward-stretching gesture of head and neck. The great crested grebe and the gannet have wing-raising and bill-fencing actions which are very similar, and the red-throated diver performs a whole series of evolutions in the water which have an obvious affinity with those of the grebe. Reference to the list of birds at the end of this volume will show that these ceremonies link together a number of Families which have been placed close to one another by the taxonomists— divers, grebes, albatrosses, petrels, pelicans, boobies, herons and storks.

Compare the foregoing accounts of albatross ceremonial with the actions of courting gulls as described by Selous (1901):

When amorous, they walk about closely together, stopping at intervals and standing face to face. Then, lowering their heads, they bring their bills into contact, either just touching, or drawing them once or twice across each other, or else grasping with and interlocking them like pigeons, raising them a little, and again depressing the heads with them thus united, as do they. After this, they toss up their heads into the air, and open and close their beaks in a manner almost too soft to be called a snap. Sometimes they will just drop their heads and raise them again quickly, without making much action with the bills. This is dalliance, and between each little bout of it the two will make little fidgety, more-awaiting steps, close about one another.

Herring gulls, like albatrosses, keep their beaks open for a perceptible time (Darling, 1938), and like them also are prone to join in when they see their neighbours displaying. It is natural to regard these similarities of display-pattern as confirmation of the close relationship between the albatrosses and the gulls.

Again, consider this account of courting wood pigeons given by Bayne (1929). The male approached the female, who made a formal kind of hop—as in a dance—which frustrated her suitor. Instead of pursuing her the cock made a similar formal hop:

This action they repeated several times alternately, and then the hen thrust her head behind her wing, which was extended slightly from her side, but not expanded. She appeared to be pecking at something and when she withdrew her head her bill was wide open in a way that suggested that she was holding some large invisible object. This whole measure was also duplicated by the male and then they repeated it alternately three times. Then they returned to the hopping and alternated this with the other measure until they were interrupted by two pigeons which had been attracted by the performance and apparently wished to participate in it. The male drove them off and returned to his mate, but by this time she had cooled and would have no more of it.

There are sufficient points of similarity in this display to suggest the possibility that the pigeons and albatrosses should be placed closer together than is the case in Wetmore's classification. Various other species of pigeon perform in much the same way (Whitman, 1919). It would, of course, be a mistake to attach crucial taxonomic importance to data of this kind, selected only to make vivid the interesting similarities in ceremonial amongst birds of different species, genera and families. Obviously, to attain reliable results it is necessary to bring a wide range of instinctive and ritual actions under review and compare them systematically. The fact that, as we shall see when we consider arena displays, convergent evolution in display-patterns may occur, shows that he who would study phylogenetic relationships must walk warily. None the less, it is evident that the systematist can as little afford to neglect homology in behaviour as in structure, and even comparisons of a most general kind such as we have been making may afford hints of taxonomic significance. Ornithologists are increasingly devoting their attention to the comparative study of behaviour-patterns and such work as that of Wachs (1933) on gulls and terns, Stonor (1936) on the birds of paradise and Miller (1937) on the shrikes indicates that here we have a field in which a plentiful harvest awaits the garnering.

As it happens, the ceremonial of pigeons and doves has been analysed in considerable detail. To Whitman (1919) belongs the credit of having shown by means of data obtained from this group that instinctive behaviour is phylogenetically constant. While no single distinguishing morphological feature common to all the pigeons and doves can be pointed out, they have, as he remarked and Lorenz (1939) has reminded us, an idiosyncrasy of behaviour which distinguishes them from all other birds except their companions in the Columbiformes, the sand-grouse (Pteroclididae), namely, the habit of drinking by pumping up water by peristalsis of the oesophagus. Thus a single behaviour-pattern indicates affinities in a large and varied group and shows that the sand-grouse, which drink in the same way, are closely related to the pigeons and doves. More recently Gifford (1941) has adduced further behavioural evidence supporting the case for the revision of the taxonomical grouping of the pigeons and doves. He argues, for example, that the tail movements while cooing and other actions of the red-throated ground dove justify its removal from the Geopeliinae in which Salvadori placed it.

The ducks have been more carefully studied from the point of view of the taxonomic significance of behaviour-patterns than any other group. Heinroth (1911) worked out a classification of the various species which was at variance with the accepted system, but Poll (1911), by studying the relative fertility of hybrids and drawing conclusions as to blood relationship, was able to confirm Heinroth's findings in every single point in which they differed from the accepted classification. Delacour (1938), making much use of his observations of living ducks in captivity as well as wild, drew up a systematic list more elaborate than Heinroth's, and recently Lorenz (1939) has carried the study of behavioural characteristics amongst the ducks still further, selecting sixteen instinctive courtship actions which are homologous between most species. By considering two characteristic actions, the 'chin lift'—a vestigial form of symbolic drinking—and a threat-over-the-shoulder gesture, he shows that the gadwall should be placed between the mallards and the wigeons. Confirmation of his theories was achieved by discovering that a pair of *Anas-Mareca sibilatrix* hybrids showed exactly the same mingling of behaviour-patterns as he had already noticed in the gadwall as a phylogenetic differentiation. He claims to have proved that Hartert's classification of the ducks is untenable.

Lorenz also draws attention to the evidence available from the study of young birds which manifest in their ontogeny the transitory appearance of phylogenetically ancient characters. Certain species of sparrow which when adult differ from the majority of their relatives by hopping rather than walking, hop when at the nestling stage. He quotes Ahlquist's (1937) observations on young gulls to the effect that in the genus *Stercorarius*—which has evidently evolved from the older gull type—the chicks in their play recapitulate the fishing methods of the genus *Larus*. A parallel is found in the fact that amongst human beings children's games often reproduce elements of the forgotten religious rites of their ancestors.

The anthropologist's approach to homologous customs in-deed provides a precedent for the application of a like technique to systematics, and parallel means of study should be fertile in results. When we find that a particularly complicated and arbitrary type of steering-gear now in use on an Indian river closely resembles that used on the ships of ancient Egypt we have reason to believe that it did in fact originate there; similarly, the eye which was painted on Egyptian ships now appears on boats from the Mediterranean to China, and from this we can deduce relationship between the various types of these ships in use to-day (Forde, 1927).

It can be shown that wherever the plough was introduced a complicated ritual accompanied it. The head of the clan, tribe or state inaugurated the agricultural year by ceremonially ploughing three, or a multiple of three, furrows, and at the same time performed other fertility-giving rites. In the course of time, as cultures diverged, elements in the ritual, such as the costume of those who performed it, came to differ very much, although in places as far apart as Greece and Siam, Yugo-Slavia and China the ritual remained substantially the same. Affinities may be detected between local versions of the ceremonial; they can also be compared with ancient Greek and Chinese records and traced back historically so that there can be no doubt that like all the doves or ducks they came from a common stem (Armstrong, in the press). If we do not press the analogy too far, it is permissible to think of the plough in its various forms as corresponding to the structure of the bird, the ritual to its display and the variations in the ceremonial trappings to the evolution of colours and adornments used by birds in their posturing.

Just as the tracing of furrows is older than the costume of the ploughman, so the mode of behaviour in displaying striking features of plumage, such as crests and the like, may usually be assumed to be more ancient than the secondary sexual characteristics which are displayed (Lorenz, 1937). Putting it crudely, the bird does not bow because it has a crest, but having the habit of bowing it developed the crest, or whatever the adornment concerned in the particular posture may be. The European night heron and the South American boatbill, an aberrant night heron, have exactly similar 'appeasing' ceremonies, a fact which suggests that the ceremony antedates the separation of the genus *Cochlearius* from the other night herons, and, indeed, is older than the crest used in it (Lorenz, 1938).

A discussion of the relationship between structure and function would take us aside from our main theme, though it might be interesting to speculate on such matters as the connexion between the flattening of the skull in certain of the Paradiseidae and the display of the head-plumes (Stonor, 1938, 1940 *a*), or whether the difference in skull formation between male and female of the extinct huia had any reference to the mode of posturing; but Lack (1941 *b*) is justified in contending that Lorenz and Tinbergen have carried too far their thesis that display movements always precede the evolution of the adornment displayed. There is reciprocity between the elaboration of movement and adornment in that once an adornment has been evolved its presence is likely to result in the intensification of the movement which displays it.

CHAPTER VI

DISABLEMENT REACTIONS.
I. TRANCE STATES

Death and 'shamming dead'—Cataleptic and paralytic states—Examples
of 'freezing' amongst birds associated with consciousness and catalepsy—
Human analogies—Ecstasy and trance during display amongst birds and
insects—Comparison between swoon states, hypnosis and 'feigning death'.

W E have already noticed that a bird's emotional responses are subject to disturbances of various kinds.
Surprise, shock, or mistaken identification of another
bird's signal may elicit inappropriate reactions and cause emotion to be discharged in unsuitable ways. We have also seen that
aimless or confused reactions may evolve into advantageous
behaviour, as, for example, when fidgeting with grass or pebbles
under the stress of emotion develops into nest-building. It may
be found that what is commonly known as 'injury-feigning'
has a somewhat similar psychological history, but before considering such behaviour let us review a series of trance or pseudo-
trance states, some of which are occasionally associated with
'injury-feigning', and speculate as to their nature and origin.

The extreme case is when a bird expires in consequence of its
terror, like the tinamou mentioned by Hudson (1892) which
dropped dead when chased by gaucho boys. In human society
we have analogous instances such as that of the man who, in a
snowstorm, rode across what he took to be a frozen plain, but
fell dead with fright when told by the innkeeper that he had
crossed the Lake of Constance (Koffka, 1935). Whether death
was due to heart failure or not is beside the point, as it is the
psychological aspect with which we are at present concerned.

Next to this we may list those states described for want of better terms as 'shamming dead' and 'cataleptic fits', in which a bird or beast becomes inert and apparently insensitive to pain. The tinamou assumes this state, and only if the shock is too strong does the bird succumb. The corncrake also, when frightened, commonly lapses into this condition, and so does the water rail. Vincent (1938) approached one of the latter which had been caught in a rat trap. It raised both wings above its head until the primaries touched, stood rigid for a moment, then closing its wings, fell on its side inert. When the cage was tilted so that the bird slid out it ran off as soon as its feet touched the grass. Cott (1940) observed a nightjar raise its wings in like manner just before 'injury-feigning'. The attitude also occurs in courtship displays. Hudson gives instances of a jaguar which was so paralysed with fear that it allowed itself to be burnt alive, and of foxes which were tortured when 'shamming dead' without evincing any sign of pain. It is well known that many insects, such as beetles and spiders, adopt the semblance of death when disturbed.

To what extent a bird is conscious when in one of these states is exceedingly difficult to say; indeed, the types of trance state vary so much and the gradations between them are so ill-defined that each instance would need to be considered on its merits. A hypnotised hen will try to peck at grains of corn dropped in front of her and may come to herself in the effort to swallow them. According to Hudson (1892) the death-feigning fox of the Argentine awakens very warily—first peeping out of his slightly opened eyes and only running away when his enemies are at a safe distance. But often the return to complete consciousness seems to be very rapid. The tinamou, for example, is ready to fly away as soon as the hand-grip is relaxed. The late Riley Fortune told me of a barn owl which went into a stiff condition with eyes closed when he took it in his hand. If he then left the room and came back in a few minutes the bird would be in its normal state. Wishing to see exactly what happened, he laid the rigid owl on a table and banged the door as if he had gone out. After a minute or two it looked around and then rose to its feet.

Coward (1926) was in the Shetlands when he watched an interesting performance. He was close to a golden plover's nest when one of the birds started 'injury-feigning', gradually retiring to a moss-hag where it suddenly collapsed with outspread wings and tail, and with its head hanging down lifelessly over a small hillock. The bird lay as if dead, but Coward made no attempt to approach. It then raised its head, squealed as if in pain —and 'died' again. Still he did not advance, so the bird suddenly came to life and ran towards him; then, with drooping wings and spread tail it tried to induce him to follow it. Thus we see that the cataleptic condition may be associated with 'injury-feigning'. Another instance is provided by a writer of almost a century ago (Couch, 1847). A man caught a skylark in a butterfly net and held her rather roughly in his hand for some little time. She lay limp and motionless, as if dead. He then threw the body away. Her captor records: 'She fell to the ground like a stone, and there she lay for me to push her about with my foot...presently the bird began moving, and with one wing trailing along the ground, and shuffling along as if one of her legs had been broken, she proceeded for a considerable distance and then took wing.'

This instance would suggest that 'injury-feigning' is a halfway house on the road to a trance condition. The behaviour of a frightened eider drake supports this. Perry (1938) writes:

I came upon an eider drake sitting out on a reef of the bay. At my distant approach he began to waddle off to the sea, but soon sank down into the seaweed. He never moved, even when I stooped to pick him up, marvelling that the eyes hidden in the black cap were dark blue. The while

I held him, his fine head gradually fell forward, as if he were dead; and when I put him in a rock pool he lay stretched out along the water. . . . He recovered slowly, paddling feebly away at first, and then more strongly, as the paralysis of fear began to wear off. When I stood up to watch his progress, he showed signs of alarm, as if aware, again, of my inherently dangerous presence. Hopping out of the pool, he waddled over the slippery seaweed of the crevaced rocks, falling occasionally, in his hurry, and almost flew into the sea, where, paddling strongly, he dived through the surf and flapped his short wings three times; in full health once more.

In like manner the rabbit waits for the stoat to kill it and the frog dances back and forth in front of the snake. An instance of the petrifying effect of fear is given by Thomson (1932). Driving a motor car he came suddenly on a pony which stood quite rigid in the road so that it had to be removed by lifting it bodily as if it were a wooden hobby-horse. A South African friend tells me of a similar experience with a duiker. When he lifted it away from the car lights it lost its stiffness and bounded off.

Less extreme instances of inhibitions caused by fear are recorded by Hudson. The black-necked swan of the pampas and the silverbill may be rendered incapable of flight by frightening them. Perhaps related to this behaviour is the common practice, especially amongst young nidifugous birds, of crouching immobile when the parents utter the warning note. An instance is on record of a young lapwing, surprised when crossing a stream, crouching under water until rescued from imminent peril of drowning, and young terns and oystercatchers will 'freeze' until the incoming tide washes over them (Armstrong, 1940). That there is no conscious choice of a squatting place harmonising with the coloration of the bird is shown not only by the conspicuous places in which they are often found, but by the fact that it is possible to turn the young bird upside down with its white under-plumage uppermost without any alteration in its condition. Young lapwings and little terns when inverted will remain motionless for as long as fifteen minutes, and young skimmers lie in the hand with head limply hanging down as though dead. Stone curlew chicks can be picked up and laid across two fingers without relaxing the rigidity of their crouching attitude. The reader will find a discussion of this kind of behaviour in the writings of Tomlinson (1930), Peitzmeier (1936) and Steiniger (1937).

It would seem possible, however, that in such cases consciousness is retained if we may judge by Hudson's (1920) experience with a variegated bittern—or heron, as he calls it. He fired at the bird and then sought it out in a bed of bulrushes.

After vainly searching and re-searching through the rushes for a quarter of an hour I gave over the quest in great disgust and bewilderment, and after reloading, was just turning to go when, behold! there stood my heron on a rush, no more than eight inches from, and on a level with, my knees. He was perched, the body erect, and the point of the tail touching the rush grasped by its feet; the long slender tapering neck was held stiff, straight and vertically; and the head and beak, instead of being carried obliquely, were also pointing up. There was not, from his feet to the tip of his beak, a perceptible curve or inequality, but the whole was the figure (the exact counterpart) of a straight tapering rush: the loose plumage arranged to fill the inequalities, and the wings pressed into the hollow sides, made it impossible to see where the body ended and the neck began, or to distinguish head from neck or beak from head. This was, of course, a front view; and the entire under surface of the bird was thus displayed, all of a uniform dull yellow, like that of a faded rush. I regarded the bird wonderingly for some time; but not the least motion did it make. I thought it was wounded or paralysed with fear, and, placing my hand on the point of its beak, forced the head down till it touched the back; when I withdrew my hand up flew the head, like a steel spring, to its first position. I repeated the experiment many times with the same result, the very eyes of the bird appearing all the time rigid and unwinking, like those of a creature in a fit.

When Hudson walked round the bird he found that it revolved slowly, maintaining the edge of the blade-like body towards him and the striped back and broad dark-coloured sides concealed. When it could twist itself no farther it whirled round with great rapidity, instantly presenting the same aspect as before.

That consciousness is not in abeyance during these 'frozen' states is suggested by the behaviour of the American bittern. Barrows (1913) saw a bird alight in a lily pond and adopt the usual erect and rigid posture when he approached:

As we stood admiring the bird and his sublime confidence in his invisibility, a light breeze ruffled the surface of the previously calm water and set the cattail flags rustling and nodding as it passed. Instantly the bittern began to sway gently from side to side with an undulating motion which was most pronounced in the neck but was participated in by the body and even the legs. So obvious was the motion that it was impossible to overlook it, yet when the breeze subsided and the flags became motionless the bird stood as rigid as before and left us wondering whether after all our eyes might not have deceived us.... This was repeated again and again, and when after ten or fifteen minutes we went back to our work, the bird was still standing near the same spot and in the same rigid position although by almost imperceptible steps it had moved a yard or more from its original station.

A remarkable case is on record of protective freezing of this kind being followed by a trance state. In a letter to Bent (1940) Kennard wrote:

I had just been investigating a big highbush blueberry bush, looking for a nest, when I discovered, to my surprise, a fledgling black-billed cuckoo, squatting on a twig about 6 feet from the ground. The little bird, which really was not able to fly, was squatting on a limb, just as little birds ordinarily do; his wing feathers were fairly well developed, but his tail was only a quarter of an inch long. When I parted the branches a trifle, so that we could see him better, and finding out that he was discovered, he promptly assumed an almost perpendicular position, with his neck stretched out almost unbelievably and his bill almost straight in the air; and there he sat, immovable, with his bill in the air like a bittern, only oscillating a trifle when the branch on which he was sitting was disturbed a little by the breeze.

My youngest son, Jack, being interested in the peculiarities of cuckoos' feet, attempted to pick him off the limb; the little bird fluttered to the ground, where he picked him up. When we had duly examined and discussed the arrangement of his toes, Jack endeavoured to put the little fellow back exactly where he had been when we had first disturbed him. Then, as he endeavoured to replace him on the limb, he suddenly went limp and, apparently, passed out in his hand, frightened to death, as I supposed. He was perfectly limp and my impression is that his eyes were closed. Jack finally, in trying to get him to stay on the limb, hung him across the limb by the neck with his head across one side and his body down the other side. Just then there came a little breeze, the body dropped, and that little bird simply scuttled in under the ferns. It was the most astonishing performance that I ever witnessed, first the stake-driver attitude, as a protective position, and then playing dead.

Most types of disablement behaviour amongst birds may be paralleled in human life. Any psychotherapist could furnish instances of men and women behaving in much the same way as the birds mentioned in these examples. A sudden fright or a startling crisis may cause a man to lose his powers of judgement and action, and to become for the time being incapable of rational activity or hysterical, so that his movements are impeded or unco-ordinated. The neurotic cases of war-time show us what happens when powerful impulses conflict. The individual develops an inhibition, amnesia, or other form of paralysis—Nature's solution of the problem of an irresistible force meeting an immovable object. Common speech recognises this state of affairs in such phrases as 'struck dumb with amazement', 'rooted to the spot with fear', and 'petrified with terror'. Brown (1940) gives case histories in which three shell-shocked patients respectively exhibited symptoms of a trance-like state, petrifaction with terror, and paraplegia; each of these showing similarity to some of the phenomena amongst birds which have been mentioned. The paraplegia patient found that his inability to walk or stand came on after an ammunition dump of which he had partial charge blew up. *Mutatis mutandis* a frightened bird might be expected to have the appearance of 'injury-feigning' as described in later pages.

If terror may induce trance states, so also may sexual emotion, though naturally of a rather different kind. It sometimes happens that in the ecstasy of courtship a bird may become so oblivious of everything else that it is possible to walk up and catch it. The male ostrich pursues the hen a number of times and then, to complete his courtship flops to the ground with his head thrown over his back and rolls from side to side displaying his magnificent plumes. At such times he is so rapt in sexual frenzy that it is possible to seize him by the neck (Selous, 1901). Schjelderup-Ebbe (1923 *b*) records that if an unmated mallard succeeds in treading a paired female no amount of ill-treatment from the bird's mate will make him desist, once he is in the 'pairing-trance'. The lesser superb bird of paradise executes a display of marvellous beauty; he spreads out a screen of velvet-black feathers, and across his chest appears a wing-shaped design of intense green: 'Two green spots of feathers over the eyes give the remarkable effect of a false pair of eyes, but in reality the bird itself has its eyes closed and seems to be intoxicated with an emotion no one could describe' (Fisher and Shaw, 1938).

According to Goodfellow (1910) the lesser bird of paradise is in such a state of excitement during the display that he is not disturbed even by the report of a gun. When displaying, Prince Rudolph's paradise bird is 'almost impossible to disturb' (Crandall, 1932). My own observations of various species of these birds in the London Zoological Gardens confirm the impression that during their display they are in a trance-like state. At a certain stage in his posturing the species named after Wallace raises himself slowly, stretching his legs as if the muscles were almost unbearably tense, while the claws scratch on the perch, so fierce is their grip. The behaviour of the bird during this paroxysm strongly suggested to my mind characteristic symptoms of an epileptic in a fit.

The capercaillie goes into the throes of ecstasy as he crows:

He begins his play with a call something resembling the word *peller*, *peller*, *peller*; these sounds he repeats at first at some intervals; but as he proceeds they increase in rapidity, until at last, and after perhaps the lapse of a minute or so, he makes a sort of gulp in his throat, and finishes by drawing in his

PLATE VII

page 72

LESSER SUPERB BIRD OF PARADISE IN DISPLAY

pages 42, 139

RIFLE BIRD IN DISPLAY

PLATE VIII

pages 72, 93

THE DISPLAY OF THE NORTH AFRICAN OSTRICH

page 90

NIGHTHAWK 'INJURY-FEIGNING'

breath. During the continuance of this latter process, which only lasts a few seconds, the head of the Capercali is thrown up, his eyes are partially closed, and his whole appearance would denote that he is worked up into an agony of passion. (Lloyd, 1885.)

Cunningham (1913) relates how he was able to catch a fighting capercaillie in his hands. It was several minutes before its adversary realised there was a stranger near 'and solemnly stalked off'. Russians call the bird *gluchar*, 'the deaf one', because while crowing it is oblivious to the noise of the hunter's approach. The argus pheasant is also said to be deaf during the utterance of its call (Beebe, 1922). Displaying or copulating penguins completely ignore a man's loud shouts uttered only a few yards from them (Roberts, 1940a). Talking and lighting a pipe do not frighten great snipe at their *Knebberplatz* (Rohweder, 1891). Selous (1909–10) noted that the 'frenzy' of the blackcock at the *lek* was sexual rather than combative. On their tournament grounds two prairie sharp-tailed grouse will engage in a kind of drumming duel with tightly closed eyes for a minute at a time as if in a trance. Then one peeps at the other, resumes his ordinary appearance and steals away, leaving the other to wake up and find himself alone. When absorbed thus in their courtship antics they fall easy prey to the crafty coyote (Cameron, 1907). The domestic cock also closes his eyes when crowing, as Chaucer noticed:

> This Chauntécleer stood hye upon his toos,
> Strecchynge his nekke, and heeld hise eyen cloos,
> And gan to crowe loudé for the nones.
>
> *The Nonnes Preestes Tale.*

Richmond (1939) says of the displaying goosander, 'In its contortions the bird appears to be oblivious to everything: it is seized and controlled by an overmastering passion.' Posturing red-breasted mergansers could be described in similar terms, according to my own observations.

In the courtship of certain spiders trance states occur. The female *Agelina labyrinthica* relapses into a cataleptic condition as a result of the male's attentions and is then carried about in his jaws grasped by a leg; she does not awaken until the whole proceedings are over (Savory, 1928). The male *Dysdera erythrina* caresses the female with his second pair of legs; at first she receives his solicitations with an aggressive demeanour and gaping jaws, but as he proceeds she assumes an hypnotic state (Berland, 1927).

There seems to be a strong similarity between the swoon-like states of birds and conditions under hypnosis (Kretschmer, 1926). It is well known that in the hypnotic trance the subject may be rendered temporarily insensitive to sufferings. The dervishes of North Africa, and similar folk elsewhere, can throw themselves into a wild ecstatic fit such that they are able to pierce their cheeks with skewers without feeling pain or drawing blood. Some birds can be hypnotised, or reduced to a state which is very similar; the Kaffirs on the Zambesi take a fowl and place it breast downwards with the head under a wing, pat the ground beside it and thus reduce it to immobility. The method succeeds up to a point with various other birds (Selous, 1927). A fowl may also be mesmerised by holding it in front of a line drawn on the ground or by seizing it and with a sudden turn laying it on its back and withdrawing the hands gradually (Katz, 1937).[1]

In the seventeenth century Father Athanasius Kircher described how, in his *Experimentum mirabile*, he bound a hen's feet and laid it on the floor in front of a chalk line so that it remained a long time as if paralysed; but it is doubtful if the line is necessary for the effect. The first account of such experiments is in Daniel Schwenter's *Deliciae Physicomathematicae*, published in 1636. But probably the mesmerising of birds was

[1] For illustrations of hypnotised birds cf. Katz (1937) and Chapman (1938).

folklore long before any account of it appeared in print; Fabre and his school-fellows used to catch their neighbours' turkeys, tuck their heads under their wings, wave them up and down until they 'went over'. Then they could be used for various practical jokes. Many animals may be hypnotised by attaching them to a board and suddenly reversing it. Stroking certain reptiles on the belly readily induces a helpless state. Large alligators succumb to treatment of this kind as readily as the small 'horned toad', or rather lizard (*Phrynosoma*), of the Californian deserts. Frogs will go into the hypnotic condition when held in one's hands, and an angry cobra will become as stiff as a stick if gripped behind the head and subjected to a little constraint. Probably the 'sticks' which Pharaoh's wizards turned into snakes had originally been immobilised by such treatment (Exodus vii, 12). A few years ago Dr Thoma hypnotised a chimpanzee at the London Zoo by fixing his attention on a bright metal knob.

The substantial identity between human and animal hypnosis has been demonstrated. It seems that catalepsy and 'feigning death' cannot be regarded as different in kind from experimentally induced hypnosis. Moreover, cats and dogs pass readily from the trance state into true sleep, from which they can easily be awakened. The experiments on dogs devised by Pavlov (1927, 1929) showed that 'Internal inhibition during the alert state is nothing but a scattered sleep...and sleep itself is nothing but internal inhibition which is widely irradiated.' Boredom induced by a tediously reiterated stimulus was found to be a protective device due to activity in the highest part of the brain. In a few instances the dog on which experiments were being made sank into a trance. It is highly probable that there is a relationship between sleep, catalepsy and 'death-feigning', trance states, hypnosis experimentally induced, and hypnosis induced by suggestion.

CHAPTER VII

DISABLEMENT REACTIONS.
II. 'INJURY-FEIGNING'

Examples of biologically useful and useless 'injury-feigning'—Communal 'injury-feigning'—Behaviour due to thwarted broodiness—Appearance and disappearance of 'injury-feigning'—Catalepsy rarely associated with breeding—'Injury-feigning' a modified form of wing-fluttering—Similarity between some courtship displays and the 'broken-wing trick'—The theory of an undifferentiated potential of emotion—The conflict between self-preservatory and reproductive drives responsible for disablement behaviour—The nature of instinct—Examples of 'injury-feigning' deceiving animals.

IN the previous chapter we have been concerned with trance states of various kinds, but instances cited there have shown that these, *i.e.* 'freezing' and 'injury-feigning', may occur in close association. When an attempt is made to analyse the circumstances which occasion the 'broken-wing trick' we find that this behaviour occurs either when there is conflict of emotions and impulses or when a strong impulse or behaviour-pattern is thwarted. Let us now review a number of examples and then consider how they may be interpreted.

An American naturalist (Saunders, 1937) was out walking by a lake shore when he noticed a wood duck with nine young on the water. Just then they were attacked by a red-shouldered hawk. The ducklings scattered and their mother immediately stretched out her head and neck, turned on one side, flapped a wing in the air and paddled in circles as though she were crippled. The hawk turned and stooped, only to collide with some bushes under which the duck had managed to find shelter in the nick of time. Thwarted of its prey the hawk flapped off to an adjacent tree-stump and perched there until the duck reappeared. Again it struck and was foiled as before. This happened four or five times. Once a couple of young ducks left their hiding-place and scurried across the water in full view, but the hawk was too interested in their frantically flapping mother to pay any attention to them. At length it flew off, having caught sight of the observer. The duck, resuming her normal appearance, called her brood together and cruised peacefully away. In all probability she had saved one of them from death.

Such behaviour has caused writers from Aristotle (*H.A.* IX. 8) onwards to attribute to birds a high measure of craftiness as well as maternal solicitude. A modern example appears in *The Testament of Beauty* by Robert Bridges:

> It is pretty to mark
> a partridge, when she hath first led forth her brood to run
> among the grass-tussocks or hay-stubbles of June,
> if man or beast approach them, how to usurp regard
> she counterfeiteth the terror of a wounded bird
> draggling a broken wing, and noisily enticeth
> or provoketh the foe to follow her in a vain chase;
> nor wil she desist from the ruse of her courage
> to effect her own escape in loud masterful flight,
> untill she hav far decoy'd hunter or blundering hoof
> from where she has bid her little ones to scatter and hide.

But in contrast to the apparent subtlety of the duck and partridge consider this incident: When Haviland (1917) was collecting and photographing on the banks of the Yenesei one day she saw four dotterel, an old bird and three young, feeding together on the slope of a hill. Over its summit came a rough-legged buzzard and at once the three young birds flew away vigorously. But the parent—presumably the male—collapsed sideways and fluttered, as if in pain, over the tundra. At first the buzzard took no notice, but as the antics continued she became interested and prepared to pounce, but Haviland fired a shot and drove the bird of prey away. She comments, 'One bird offered to sacrifice itself for its young, who did not need it, to another who did not expect it.'

The fact is that birds frequently go into these semi-paralytic states at times and in places where they can serve no useful purpose. Abel Chapman (1897) flushed a black-throated diver from her eggs which were just chipping, but after flying thirty yards the bird fell heavily on the water, to all appearances with a broken wing. For several seconds she lay flapping helplessly on her side, swimming in narrow circles as if paralysed; but half an hour later the same bird was seen flying vigorously three hundred feet in the air. The only possible enemy which might thus be decoyed from the nest into the water is the otter, and it is highly improbable that the 'device' was evolved in regard to this one animal. Moreover, such behaviour by black-throated divers is rare. The incident may be compared with Selous's (1905 a) account of a waterhen which he saw 'fall into a sort of convulsion on the water' when surprised, and Lowe's (1934) description of a Kentish plover stopping dead in flight and falling 'as if it had struck a wire'.

The collapse of birds in flight is so gratuitous as a contribution to the deceptive appearance of the performance that it may be considered an indication of the 'paralytic' nature of the reaction; and this is reinforced by observations such as Lack's (1932) on the nightjar's 'impeded' flight. This bird does not show disablement behaviour unless flushed from the eggs or young—suggesting that it is due to the conflict of incompatible drives.

Like many other types of bird behaviour 'injury-feigning' may become a communal performance, though it is difficult to believe that the simultaneous collapse of a number of birds serves any useful purpose. Dewar (1913) describes such a display as he witnessed it in India on an islet in the Gogra river:

A search of less than a minute served to reveal a couple of eggs placed on the bare ground between two small plants that were growing out of the sand. As I stooped down to examine these eggs I looked round and saw a very curious and pretty sight. Swallow-plovers were surrounding me. They were nearly all on the ground and striking strange attitudes. Some were lying on the sand as though they had been wounded and fallen on the ground; others were floundering in the sand as if in pain; some were fluttering along with one wing stretched out limply, looking as though it were broken; while others appeared to have both wings broken. I did not count the birds, but at least twenty of them were seemingly wounded.

Seebohm (1885) describes as a usual experience seeing a dozen or so of these little pratincoles 'feigning lameness' before any one of them had laid an egg. The fact that this behaviour may occur either before or after there can be biological advantage in it suggests that like so many behaviour-patterns associated with the breeding cycle it matures as an innate inherited mechanism. The pied stilts of New Zealand also nest in colonies and collapse communally. When disturbed they attract attention by all kinds of peculiar antics, then, according to Guthrie-Smith (1925), 'The end comes slowly, surely, a miserable flurry and scraping, the dying stilt, however, even *in articulo mortis*, contriving to avoid inconvenient stones—upon which decently to expire. When on some shingle beach well removed from the eggs and nests half a dozen stilts—for they often die in companies—go through these performances, agonising and fainting, the sight is quaint indeed.' If a pair of Australian white-fronted chats are alarmed at their nest, not only will they give a disablement display but neighbouring chats, attracted by the fuss, join in the demonstration (Chisholm, 1934).

In most of these instances the behaviour can be considered as due to a conflict between fear and the impulse to brood or tend the chicks, but in some instances fear seems to play an unimportant part and 'injury-feigning' appears to be the outcome of thwarted broodiness. Black-necked stilts disturbed while brooding alternate 'injury-feigning' with squatting as if on eggs (Lack, 1941 *b*). An ornithologist (Grimes, 1936) discovered a nest of Swainson's warbler; the bird was so tame that she would not leave the nest until pushed off, and even then, when a hand was held over the nest, straddled the fingers in trying to get on to brood! When driven off she went fluttering along the ground in the manner of a crippled bird. This warbler was not badly frightened, for in a few minutes she was back on the nest accepting proffered flies from the naturalist's fingers. A somewhat similar happening is recorded by the same writer: A Florida prairie warbler fed her young without concern whilst a camera was set up close to the nest, but when the chicks, being alarmed, sprang from the nest, she fluttered to the ground and grovelled on the sand in what appeared to be a kind of frenzy. This lasted but a moment and the bird quietly set about feeding the fledglings as soon as they were settled in their new surroundings. The dove of the Galapagos Islands flutters painfully away, although the birds of the islands as a whole are unafraid of man (Swarth, 1935), and the Kerguelen Land teal cannot be induced to fly far, but if surprised at the nest flaps about as if maimed, despite the fact that there are no four-footed or human enemies in this region to decoy away; though possibly the device may deceive the skuas which prey on the young (Moseley, 1879). The birds were so confiding that a procession of them came to meet the naturalist of the Challenger Expedition. Such instances as these show how fantastic was Hudson's theory that the 'broken-wing trick' is due to birds being incapacitated by pain.

In a number of species it has been noticed that 'injury-feigning' either increases daily as incubation advances or is correlated with particular phases in the breeding cycle. The killdeer plover accentuates its disablement reactions each day of incubation (Friedmann, 1934), and according to Hume (1890) the little pratincole of India performs in a most realistic manner when there are hard-set eggs or young, but is quite unconcerned when freshly laid eggs are being stolen. This, however, does not agree with Seebohm's experiences. The female bay-breasted warbler is somewhat exceptional as she 'injury-feigns' when incubating but not after the young have hatched (Mendall, 1937). Saxby

(1874) says that the dunlin 'pretends' to be maimed when her eggs are nearly hatching and not when they are newly laid. Parents with young do not act in this extreme way but run tamely around the intruder. It is not surprising that 'injury-feigning' should develop, reach a climax of intensity and then disappear, for there are many other modes of avian behaviour of which this is true. Sequences and patterns indeed rule the lives of birds. Amongst some young birds the 'freezing' habit rises and wanes. For example, shortly after hatching young oystercatchers crouch but soon start moving again, and they struggle when seized. By the third day crouching persists, even when the chicks are handled. On the fifth day the response is almost perfect, there is marked limpness and any position but inversion is retained; but in the second week the birds do not struggle, even when placed upside down, and will remain crouching for hours. The tendency to squat disappears entirely with the acquisition of flight (Dewar, 1920 a). Fear is often tardy in manifesting itself, but makes its appearance suddenly. Herrick (1901) writes: 'The instinct of fear comes with a certain maturity of the nervous system, with comparative suddenness, but is usually timed to correspond with the development of the wing quills and the forms of flight.' He points out that the belted kingfisher has no fear of man at the age of twenty-four days, although fully feathered, but twenty-four or twenty-eight hours later it shows its terror by taking flight.

'Injury-feigning' very rarely occurs except when a bird is anxious for its nest and eggs or young, but a sparrow, chased by a rat, will sometimes flutter weakly along the ground as if under an inhibition like that which impedes a rabbit when a stoat is on its track. It is most usual when the eggs are well incubated or there are small chicks, though, as we have seen in the case of the dotterel, it may linger on, outliving its usefulness.

Few birds which do not nest on or near the ground exhibit the 'broken-wing trick', but the bulbuls and minivets might be admitted as exceptions, for they sometimes nest high up (Dewar, 1928). Various tree-nesting Australian birds, such as the yellow-tufted honey-eater, constitute exceptions (Chisholm, 1934). Eagles, falcons and hawks occasionally plunge headlong from their eyries, almost striking the ground before dashing away (Jourdain, 1936-7).

Whereas 'injury-feigning' is characteristic of the period when eggs or young are being tended, the full cataleptic reaction is rarely associated with nesting. Many birds which are particularly subject to cataleptism, such as the corncrake, never collapse when the nest is threatened. On the contrary, a corncrake will courageously attack a dog or fight a rat until it is too crippled to escape—as in an instance which came under my notice. This difference in behaviour deserves emphasis, for it shows that both these phenomena are what they are, not merely because of the paralysing effect of fright, but that they owe their survival to the selective processes of nature. Consider two birds of the same species in the far-distant past. An enemy appears at the nest in each case. Terror causes one bird to go into a trance; the other is able to flutter a little. The first bird and its progeny would be destroyed, the second bird might escape; and even if it did not, its young would have a good chance of survival through the distraction caused by the parent's struggles. Cataleptism is, so far as the race is concerned, utterly dysgenic if associated with incubation, but has survival value for the individual in many instances when no other means of escape remains.

It is significant that almost every gradation in behaviour is exemplified by agitated birds from merely fluttering the wings to remarkable simulations of injury. When I approached a short-eared owl's nest one of the birds swooped down to a hillock about twenty-five yards away, and holding its wings out from its body in a drooping position, quivered them rapidly, at the same time emitting a squealing or chirruping note. The performance reminded me strongly of the movements of a sparrow fledgling begging for food (Armstrong and Phillips, 1925). But in Orkney one of these birds alternately threatened an intruder and fluttered on the ground as if incapacitated (C. E. Baker, 1937). A male long-eared owl bringing prey for the young to the nest is greeted by his mate with quivering wings (Hosking, 1941). When young lesser whitethroats are handled the female runs about with wings and tail spread and feathers relaxed (Howard, 1907-14), whereas a willow warbler, disturbed from the nest has been seen to alight on some palings giving all the appearance of being wounded— beak slightly open, left foot trailing and right wing hanging as if the humerus were broken (Siddall, 1910).

The fluttering of drooped wings is by no means confined to occasions when a bird is frightened. It may occur in the chicks' begging movements and the adults' sexual display. Just as the anxiety display may vary from gently quivering wings to complete collapse, so may the courtship posture range from wing-fluttering or drooped wings to prostration.

It is not self-evident that there should be similarity in displays with such diverse motivation. The most reasonable interpretation of the facts is that both in courtship and 'injury-feigning' the crude emotional expression of the hungry chick appears in a modified form. In some instances we have a reversion to the primitive display, in others it has been elaborated and become part of a complex behaviour-pattern.

No meaningful gesture is more widespread amongst birds than wing-fluttering. It is so characteristic of importunate young that examples would be superfluous, but it is worth while to note that the gesture may be used by a bird on occasions of very different emotional tone. The attitudes of willow warblers of both sexes prior to coition are identical with the appearance of the female when frightened at the nest, and there is a similarity between this attitude and that of the chick craving food (Howard, 1907–14). During sexual excitement a grass-hopper warbler's movements are similar to its display of parental anxiety (ibid.). As Howard's pages show, the fluttering courtship movements of many other warblers are similar to their anxiety posturings. The expansion and quivering of one wing by the reed bunting may be associated with an attack on an intruder, the enticement of a mate or concern about the young, according to the phase of the reproductive cycle in which the bird happens to be (Howard, 1929). Compare Rickman's (1931) sketch of a frightened young woodcock with Hosking's (1940) photograph of a nervous adult and Seigne's (1936) description of the woodcock's attitude during courtship —little or no difference can be perceived. An anxious reeve runs about with wings hanging loose, making little jumps into the air rather like a ruff on his court, and the display of partly opened, waving wings is employed by young Galapagos finches when begging food, by males to intimidate other males and in courting the female (Lack, 1941 a).

Incidentally the elevation or quivering of one wing appears in the displays of many birds, such as the purple sandpiper (Keith, 1938), buff-breasted sandpiper (Rowan, 1927 a) and skylark (Howard, 1929). Before a song sparrow fights for a territory he flutters one or both wings held up into the air and sings softly (Nice, 1937). While the female hedge sparrow quivers a wing she turns to the male with bill opened widely, as she did as a fledgling. I have noticed an Alpine accentor chick beg thus while fluttering one wing. The Kentish plover, marsh warbler (Howard, 1907–14) and satin bower bird (Gould, 1848) extend alternate wings when sexually excited. The drooping of one or both wing is usual in the nuptial display of many Gallinaceous birds. Wing-quivering is part of the sexual display of species from birds of paradise to sparrows and is also characteristic of insects such as the fly *Drosophila melanogaster* which waves a wing during courtship (Richards, 1927), and the mantis which shakes its wings with a convulsive tremor before embracing the female (Fabre, 1912).

Some species have a sexual display which is almost identical with 'injury-feigning'. The amorous rock thrush beats his wings on the ground (Hingston, 1933) and the pied wagtail has a somewhat similar display (Howard, 1929). The male Misto fieldfinch lies prone before the female with outspread, tremulous wings (Hudson, 1892). Yellow buntings (Turner, 1924) and wheatears (Lloyd, 1933) fall prostrate when courting, while the ecstatic sprawlings of a love-sick guinea fowl suggest the paroxysms of a stricken bird. Lack (1932), writing of the night-jar's 'injury-feigning', remarks that actions unmistakably associated with courtship at times replace the more typical behaviour. The 'impeded' flight of this bird is reminiscent of courtship flights in other species.

The fact that every stage can be found from simple wing-quivering to elaborate sexual display and 'injury-feigning' suggests that both types of behaviour have been built up from the begging movements of chicks—further confirmation of our theory that many display-patterns have evolved from attitudes of the hungry young. Emotion thus early manifests itself in ways which are employed in later life to express craving for the satisfaction of other requirements and excitement in other situations. As the gesture may become more specific in the later courtship display so too may its emotional import. The theory which can best explain all these facts is that there exists in birds an undifferentiated store of emotional energy or *libido*, seeking to express itself, but sometimes unable to do so adequately, and at other times diverted into inappropriate channels.

If this theory be sound, it tends to support some of Freud's general principles, but also to suggest modification of his theory of sex as the fundamental element in the psychology of living creatures.[1] Conation is first shown in desire for food and warmth. It later becomes specialised at certain seasons in sex desire, and uses for its expression the movements which were originally associated with the desire for food.

The cause of disablement reactions—conflicting or thwarted impulses—should not be confused with the origin of their mode of expression in the wing-quivering of the chick. 'Injury-feigning' is a reversion to a pre-fledging type of behaviour, a comparatively simple pattern from which elaborate displays have been evolved. The more highly organised patterns break down when the bird is confronted with a situation to which it cannot make adequate response. And in extreme cases there may be such a breakdown that even this primitive pattern does not operate, and disablement, trance or collapse ensue.

The two main drives responsible for the behaviour of a bird are self-preservation and reproduction, the one permanent, the other cyclic. Within the ambit of each of these there are systems of behaviour-patterns, some of which are associated with certain phases in the life cycle, such as migration, courtship and brooding; and within these are minor rhythms, as for example incubation periods with associated nest-relief ceremonies, about which I shall have more to say later. We can thus conceive of a psychic framework composed of these behaviour-patterns and dominated by the twin drives to preserve the individual's life and perpetuate the race. It is a plausible hypothesis that disablement reactions arose through the clashing of these drives, that the

[1] It has been shown that in both sexes of rat, studied when hunger and sex needs are maximal, the desire for food is greater than the drive to satisfy sex-desire (Warner, 1927, 1928; Bierens de Haan, 1940).

result was a harking back to a primitive response, and that this, having survival value, became fixed as a behaviour-pattern in its own right. We shall shortly see that it is not unusual for behaviour-patterns originally relevant to one type of behaviour, such as preening, to become elements in other behaviour-patterns, as, for example, courtship.

This theory makes it easy to understand why some forms of disablement reaction do not closely simulate the actions of an injured bird. Any movements which give the bird an odd appearance as it leaves the nest and render it conspicuous and apparently easily approachable may have value in distracting attention from the nest or young, so long as they are not carried so far as to preclude escape from the creature which excites them. Such actions, tending to the survival of the race, would not be eliminated by the forces of selection.

Incidentally, the interpretation of bird activities on the basis of systems of behaviour-patterns organised under two master-drives, whose physiological springs we are now able to explore, gives promise of providing a more exact meaning for the word 'instinct'—a term which has been avoided by ornithologists such as Howard because of its vagueness, though still used in a variety of significations by psychologists. The organisation of behaviour-patterns can be traced to the *corpus striatum* of the brain. While the cerebral cortex which in its highest refinement is responsible for reason in man remains undeveloped, the *corpus striatum* of birds is more highly evolved in size and complexity of structure than in any other living creatures. Moreover, as will be shown later, the injection of androgens (or forms of male hormone) will induce many male behaviour-patterns in female birds, such as singing, territory defence, despotism and court-ship display, as well as accentuate male behaviour in male birds. Thus, the physiological foundations of behaviour are much more fully understood than they were a decade ago, and we are coming within sight of a definition of instinct in mainly physiological terms.

Lorenz (1939) has made important suggestions in this connexion. Instincts are behaviour-patterns co-ordinated in the central nervous system which are not determined in their form by external stimuli; instinctive behaviour is endogenous automatism and is thus physiologically and causally different from all individually variable behaviour-patterns. Instinctive acts may be recognised by their conformity to the rule: The longer an instinctive act is not released, the greater is the intensity with which it occurs in response to a given stimulus-situation. Or stated in another way: The longer time has elapsed since the last release of an instinctive act, the less complete a stimulus has to be in order to release it. 'Taxes' are innate movement-patterns which are not only set in motion or 'released' but are continuously directed by external stimuli. Most reaction-patterns consist of an intermingling of instincts and taxes. 'Drives' are taxes which are dependent on a certain level of intensity of a definite instinctive act.

It is not surprising that when the two master-drives with all their associated physiological and psychological systems come into conflict, or one of these drives when dominant in a bird's life is thwarted, that the outcome is a greater or lesser degree of disorganisation in the behaviour-patterns.

Lorenz (1938) says: 'It is typical of instinctive action that in the case of two conflicting reactions being released by the same object, the resultant behaviour never shows a purposeful compromise between the two impulses, but a perfectly aimless intercalation of two actions that serve absolutely contradictory biological ends.' The reason is that behaviour-patterns are rigidly organised and correlated with the internal state. It so happens that although 'injury-feigning' is not a 'purposeful compromise' it does serve a useful purpose.

Doubt has been expressed by some naturalists as to whether 'injury-feigning' is actually effective. Familiar as I am with the 'trick' I was once deceived by a ringed plover into running after it. Whilst searching for its nest I suddenly saw a checkered, fluffy object struggling on a bank. Convinced that it was a young sheld-duck I ran to it, only to see it transform itself into the ringed plover and fly away. But the extent to which man is deceived is not a fair criterion, as he is only one of the enemies which a nesting bird has to circumvent, and on the whole not the most dangerous; for if we review the history of birds, allowing for the fact that man has only comparatively recently become a common animal and is not so to-day over vast areas of the earth—forest, marsh, desert, steppe, tundra and remote islands—we must recognise that the deceptive devices of the animal world do not owe their origin to the necessity of outwitting him. Does 'injury-feigning' deceive other animals?

Frances Pitt (1927) is one of many who provide evidence that this question must be answered in the affirmative:

That the trick is effective when practised upon such creatures as foxes and otters I can testify, for I have seen Moses, my otter, thoroughly fooled by a wild duck (mallard). This duck was of the domesticated strain known as the decoy duck, which is like the true duck in all particulars, and can fly as well as the wildest. She had made her nest about five yards from the edge of a pool, under the shelter of a small bush, and was sitting close, when I took Moses for a walk in that neighbourhood. The otter was poking about in the grass, when she appeared to wind the sitting duck, and sprang forward, making a grab at her. Now the duck was tame, very tame, and had I disturbed her would either have hissed and pecked at me or merely toddled away. She had certainly had no previous experience of an otter's attack, or had any occasion to practise the broken-wing trick, yet she now floundered from her eggs and scrambled, flapping, towards the water. The otter dashed after her, into the water, along which the duck splashed, quacking loudly. On the duck went, the otter swimming her hardest after her, until at the far end of the pool the duck took wing and flew off. The otter's air of bewilderment was laughable.... The duck's ruse was completely effective as a means of drawing the foe away from the eggs and the otter was thoroughly deceived.

Huxley (1925) doubts the efficacy of the 'trick', and referring to the action of an eider duck which he saw shuffling away with a drooping wing, describes it as 'singularly poor feigning of injury which would hardly deceive a fox', but none the less he mentions that a member of his party saw a duck decoying away a fox on the mainland of Spitsbergen by this very device. When I have seen the eider duck perform, it was much more a matter of hobbling away than simulating the appearance of a broken wing, but I am sure that such an animal as a fox or other predator would be tempted to pursue such apparently easy prey. Huxley acknowledges that the efficacy of 'injury-feigning' on the part of the purple sandpiper is very great. The bird returns and repeats the performance again and again. The knot also employs disablement reactions successfully to decoy dogs, foxes and weasels from its chicks. The Texas nighthawk 'injury-feigns' when approached by a man, but does not do so when the intruder is a quadruped (Pickwell and Smith, 1938), and some nightjars display aggressively at night but 'injury-feign' during the day. The killdeer plover flies towards an approaching man or dog and flaps about a short distance ahead, but when a horse or cow approaches it dashes suddenly in its face (Taverner, 1936).

The stone curlew 'feigns injury' with remarkable realism. Hosking (Woodward, 1938) was photographing a pair when a retriever rushed into the nesting area:

The hen saw the dog before the cock, and ran towards him from the nest. At once the dog chased her, while she lured him all across the field, then sprang into the air, circled round, and returned to her duty. As to the cock, all that he did was to rise from the eggs and walk about until the dog had vanished, when he returned to the nest.

Thompson (Ogilvie-Grant, 1896), writing of the ruffed grouse, says that he has often seen a dog duped by its 'injury-feigning' and has little doubt that a mink, skunk, raccoon, fox, coyote or wolf, would fare no better. He records that if the bird has young she will run to meet and mislead the enemy before he reaches them.

C. R. Mitchell, in *Nature's Story of the Year*, tells of a blackbird which repeatedly led a' cat away from its young by 'injury-feigning'. So often did this happen that a fox terrier learned the meaning of the bird's cries, and whenever it heard them rushed forth and took the cat in the rear. According to Jourdain (1936-7) a pheasant which Witherby disturbed brooding her chicks fluttered with drooping wings just out of his dog's reach for twenty or thirty yards, and then flew into a tree. Lynes (1910) records a lapwing decoying a stoat four hundred yards away from its unfledged young.

It would be easy to multiply such instances of the effectiveness of 'injury-feigning', but enough has been said to make it evident that however blind the gropings by which Nature reaches her ends, and however devious the paths which she follows in achieving them, she succeeds in enabling the foolish things of the world to confound the wise.

CHAPTER VIII

THE EXPRESSION OF THE EMOTIONS

Inappropriate displays—False-bathing, -feeding, -drinking, -preening and -nest-building—Premature and belated activities due to physiological causes —Activities aroused by endogenous stimulation—Song and sexual display aroused by excitement—Hingston's theory discredited—Lorenz's display-types—Bi-valent display—Animal emotional behaviour less differentiated than human—Emotion aroused when instinct is without adequate outlet— Conflict theory of emotion—Analogy between an argument and the display of birds.

THE preceding discussion has shown that ineffective, confused or inappropriate responses to stimuli are not uncommon amongst birds. Such reactions may be of many different kinds, and as they give us some insight into the nature of a bird's instincts and emotions we may consider a few of them here.

It is well known that excitement, fear or pain may provoke birds to burst into song. A stone thrown into a reed bed will start a sedge warbler singing, and a lark which has escaped from a hawk trills loudly (Kennedy, 1935). A chestnut-shouldered hangnest wounded by Hudson (1920) fell into a stream; as it floated it began to sing and continued to do so even when he took it up in his hand. Lack (1939a) mentions a trapped robin which sang for a short time while he was holding it. There are analogies in human behaviour. People who have just escaped from a railway accident or pit disaster have been known to burst into hysterical peals of laughter. According to Darwin (1872) a man in peril on a cliff near the Golden Gate, San Francisco, alternately screamed and laughed.

Sexual posturing may take place incongruously. Disturbance may cause a whole flock of avocets to start copulating (Makkink,

1936). A gull has been seen to display sexually after losing a fight with a rival, and another was observed by the same naturalist (Darling, 1938) to bob her head in a type of display which is usually mutual as she stood beside two fighting males. Mergansers and terns display immediately after the sexual rite. A pair of goldfinches, surprised with their fledged young, postured almost exactly in the same way as a courting male, jerking their bodies from side to side as if on pivots (Oldham, 1938). When an ornithologist (Keith, 1937) suddenly showed himself to two red-throated divers on their tarn they performed the 'roll-growl' and 'snake-race'—both associated with sexual situations. Amorous cock ostriches fall to the ground in an ecstatic state; according to Lack (1941 b) they display in this way when an aeroplane passes over them. Possibly, however, threat and courtship postures can be mistaken for one another (Martin, 1890). Display aroused by a wrong object rather than an inappropriate experience is in a different category. An argus pheasant courted a stone water-trough in default of a female of his own species, as the fireback pheasant in the pen would not stand still (Bierens de Haan, 1926 a), and pigeons have been known to posture to a man, a bottle and a crust of bread (Whitman, 1919). Instances of this kind akin to fixations are mentioned in Chapter XVIII.

When I frightened a black-headed gull from her hard-set eggs on a small marine island she began to bathe in an energetic, nervous way only a few feet from me. It is well known that many birds, such as ducks and grebes, bathe under the stress of sexual emotion. This 'false-bathing' is not always easy to distinguish from real bathing; indeed, the two merge into one another. Parties of red-breasted mergansers will frolic in sexual excitement on the water, splashing, diving and flapping, but as ardour diminishes they separate to bathe and preen placidly. 'False-bathing' may, indeed, become part of the sexual display-pattern. Three drake king eiders have been seen on a beach displaying to a duck, bowing heads, stretching their necks and uttering a drumming call. At frequent intervals one of the drakes waddled to the water and bathed. It seemed as if this constituted part of the ceremony (Brooks, 1915). The extraordinary topsy-turvy diving display of red-throated divers is preceded by bathing performed in a spasmodic, excited way, the wings being quivered rather than beaten on the water. Moreover, the expulsion of an intruding bird calls forth similar behaviour (Selous, 1927). I have seen whooper swans 'falsebathe' in such circumstances. After coition mute swans (and various ducks and other water birds) bathe and preen. Black guillemots also bathe in the intervals of their 'water-dances' (Armstrong, 1940).

'False-feeding' is another common type of 'substitute activity'. Larks, tits, avocets and ducks are amongst the many species in which it occurs (Tinbergen, 1939b). In the intervals of fighting for territory snow buntings appear to pick up something from the snow, but careful observation shows that they only perform the motions of picking. A male snow bunting will also 'false-feed' when refused copulation by the female (Tinbergen, 1939a). Ringed plovers dally picking amongst the stones in a way most aggravating to the photographer in a hide by the nest, but the action is due to nervousness caused by his presence. In some species actual feeding may occur, but that it is a 'substitute' activity is sometimes evident from the unsuitability of the food, as in the case of the gannet mentioned earlier which ate its nest. Nervous monkeys will 'false-feed' (Carpenter, 1934), mice introduced into a cage with strange mice will gnaw anything available, and a horse, immediately after an accident, crops grass assiduously. The 'bill-wiping' performed after sexual or other excitement by species as various as the sparrow and the great bird of paradise—to mention only

two of the many in which I have observed it—is another type
of substitute activity.

'False-drinking' is also a common phenomenon. In some
species it is not easy to say whether the gesture is most suggestive of feeding or drinking. This is true of the whooper swan;
disturbed with its cygnets, the bird constantly lowers its beak
almost to the surface of the water. Courting waterhens frequently peck at the water (Morley, 1936). For hours at a time
black guillemots dip their bills every few seconds as if they were
sipping; cormorants and shags do likewise during sexual excitement, but also to some extent out of the breeding season and
away from their nesting haunts or others of their own species
(Armstrong, 1940). The regularity of the practice, continued
over long periods of time, makes it impossible to believe that the
birds are really drinking. I have, however, seen a mute swan,
aroused from the nest, go to the sea's edge to drink, and
Murphy (1936) states that Galapagos cormorants in New York
aquarium eagerly drink salt water. Red-throated divers continually 'false-drink' on their breeding lochs.

'False-preening' is so common that instances will be remembered by everyone who has watched birds. It also occurs
in women. After an accident they will comb their hair with
exaggerated care. After 'injury-feigning' two or three times a
whitethroat will 'false-preen'. Kittiwakes preen after gulls have
stolen their chicks. Razorbills break off fights in order to indulge
in this 'habit-preening', and paired birds constantly preen each
other. Guillemots show so little territorial aggressiveness that
they preen their neighbours as they incubate (Perry, 1940). The
rough usage which the gannet gives his mate—to the extent of
denuding part of her neck of feathers—may be regarded as a
variation of false-preening. In the courtship of great crested
grebes the bill is placed amongst the wing-feathers (Huxley,
1914). False-preening behind the wing is used as a threat by the

crane but precedes copulation in domestic pigeons (Lorenz, 1939). Hence we see that a nervous habit may become formalised and associated with a specific emotion, although in itself emotionally neutral. Courtship bathing amongst king eiders may be another illustration of this, and also the guillemot's habit of bowing in emotional situations which Perry (1940) thinks has originated from the bent posture necessary in arranging the egg and feeding the young.

'False nest-building' varies from the haphazard picking up of a pebble or grass-stem to building an almost complete nest. Examples were given in Chapter II. It is occasioned by various kinds of excitement. A female yellow bunting will collect dead grass and let it fall after having driven an intruder from her territory (Howard, 1935). The male Gould's manakin terminates his court display by picking up leaves: Chapman (1935) comments, 'This act seems to be more or less sexual in character.' In the Galapagos finches this type of behaviour has courtship significance (Lack, 1941 a). So far as its psychological springs are concerned probably it does not differ from 'false-preening'. In the one case the bird fidgets with itself, in the other with another object. Endless examples could be given of similar behaviour in other animals, from the capuchin monkey shaking down leaves on the intruder and the excited dog which picks up a stick without previous training to the nervous child gnawing a pencil or playing with her dress.

Unlike the other activities with which we are here dealing the simple action of picking something up has, in a great many species, been worked up into an elaborate series of movements —nest-building—and has become saturated with emotional content of different kinds, just as in the case of preening behind the wing. For the herring gull substitute nest-building has become fraught with threat significance, and for the Adélie penguin and many other birds it is a courtship action.

A simple nervous gesture may thus acquire high specific emotional valency. Along these lines what are called 'symbolic' activities can be interpreted. An action by constant association with an emotional situation comes to 'symbolise' that situation and is able to play a part in arousing it. Rather naturally 'intention-movements' come into this category. They sometimes evolve until they have a definite character of their own.

Just because false nest-building activities may appear in a very simple or highly elaborate form they are particularly instructive. They may anticipate actual nest-building or appear belatedly. In the gannet they do both. As remarked earlier such anachronistic activities are due to the endocrine system stimulating behaviour-patterns in inexact correlation with the environment. A certain high threshold of excitement and adequate co-ordination have to be reached before the complete activity can be performed. External conditions, such as the march of the seasons, and internal factors, more particularly hormonal control, operate together to direct energy to the performance of particular actions at the optimum time. It is of the nature of instinctive acts that they may be adumbrated or represented inadequately if the excitement is insufficient. These 'intention-movements' may either be similar to the complete performance, as in fidgeting with nesting material, or different (according to the threshold of excitement), as in the flight intention-movements of wild geese (Lorenz, 1939). Certain types of sexual display are coitional intention-movements more or less ritualised.

So-called 'imaginary' nest-building, as when a gannet passes material to a non-existent mate or a humming-bird weaves non-existent material with elaborate motions into a non-existent nest (Lorenz, 1939), illustrates the opposite condition of affairs to that in which nest-building is imperfectly performed. The energy behind the instinctive act is so great that even in default of the appropriate environmental conditions the action is performed. When a caged cuckoo at the season of passage flutters its wings as in a trance (Kidd, 1921), or a guillemot feeds a non-existent chick (Perry, 1940; Johnson, 1941), or gulls enact the 'change-over' before there are eggs to incubate (Darling, 1938), or passenger pigeons take turns in brooding the eggs before they are laid (Whitman, 1919), or a young captive starling which has never caught a flying insect goes through all the motions of doing so, including killing and swallowing it, although it is not there at all (Lorenz, 1939), or a robin repeatedly attacks empty space from which a stuffed specimen has been removed (Lack, 1939a), we have other examples of elaborate behaviour-patterns released by internal or endogenous stimulation.[1] How deeply rooted are some of these movement-patterns in the physical organism is shown by their co-ordination in the central nervous system independently of receptors—not excluding proprioceptors (von Holst, 1935, 1936 a, b, 1937 a, b).

All these phenomena show that the internal condition has a highly important function in determining behaviour. Lorenz accounts for them by what he styles the 'accumulation of reaction-specific energy', but I suggest that this concept may profitably be brought into relationship with the term 'emotion'; a term which is by no means clearly defined by the psychologists and is too readily dismissed by biologists as concerned with 'unobservables'. 'Emotion' is too useful in denoting tersely certain familiar and not otherwise easily describable phenomena for it to be permissible to banish the term. Like electricity it may be difficult to define, but the phenomena to which it refers are definite enough.

We have reviewed illustrations of the ease with which in-

[1] Empid flies presented with inedible objects by the males perform the movements of eating (Richards, 1927); but probably such behaviour is of a different nature from that we are considering.

appropriate behaviour is called forth and seen how excitement may find expression in incongruous activities. How is it that feeling and its expression can be so readily disassociated? This can be understood if emotion is conceived as analogous to a store of energy within the organism; but before considering this in greater detail, let us consider what I shall call the bi-valency of some forms of bird display.

We need not dwell on the now fully established fact that song repels the male and attracts the female; Selous (1933) set it down as an observed fact in his diary in 1909; Nice (1937) and Lack (1939a), amongst others have confirmed it. In regard to posturing the situation is more complicated; there are some displays which are employed both to attract or stimulate the female and intimidate enemies. For example, I have seen a peacock use his complete sexual display very effectively to frighten a bantam out of his pen. Kirkman (1937) records that the black-headed gull's 'forward display' has both a sexual and a threat connotation. Hingston (1933) used such facts as these to bolster up an extravagant theory that all courtship display is the expression of hostility and that even 'the sex-act is not a mere male-female contact, but rather an act of fierce hostility directed for a time against rivals of the same sex and securing fulfilment through an act of union with the opposite sex'. A detailed critique of this totalitarian theory, which sees fear and enmity dominating the whole of creation, is not called for here. Hingston was mistaken in regard to many of his facts and particularly in his thesis that all bird courtship and defiance displays are identical. In some cases, indeed, they appear to be so—though this may be due to inaccurate observation or the crudity of our perceptions —but in many others they are not. He claims, for example, that the blackcock's display when fighting is the same as before the greyhen, although long ago Selous (1909–10) showed that the blackcock reserved a special tilt of the tail for showing off to the female. Lack (1939 *b*) has confirmed the distinction between the two displays and shown that the autumnal display of blackcock (and robins) is aggressive and not courtship. The male ruffed grouse, when courting the female, walks round her with head held high and tail lowered, but the much more spectacular display, when the bird with hissings and head-shakings erects his ruff and spreads his tail fanwise is used in the 'war of nerves' with other males (Allen, 1934). Similarly the great crested grebe (Huxley, 1914; Venables and Lack, 1936), various species of penguin (Roberts, 1940*a*), redshank, black-tailed godwit (Huxley and Montague, 1926), nightjar (Heinroth, 1909), nuthatch (Venables, 1938), robin (Lack, 1939*a*) and song thrush (Tinbergen, 1935) have quite distinct intimidatory and sexual displays. In many species the adornments which have a threat function are clearly defined—the robin's breast (Lack, 1939*a*), the domestic cock's neck decorations (Lorenz, 1935), the red-winged blackbird's red markings at the angle of the wings (Noble and Vogt, 1935), and probably the chaffinch's white shoulder-patch (Lack, 1941). These, according to the terminology introduced by Poulton (1890) and adapted by Huxley (1938*a*) are aposematic or threat features as distinct from sexual or epigamic. It is a commonplace that some species have threat displays which are quite different from their epigamic posturings. The sun bittern, for example, has a simple form of courtship, but confronted with an adversary the wings are thrown open and held almost vertically, the tail is erected and the intruder confronted with a startling apparition (Stonor, 1940). The common bittern when in an attitude of menace raises its crest and spreads its neck feathers (Portielje, 1926).

In those species in which aposematic and epigamic posturings are similar the situation can be interpreted to some extent according to Lorenz's (1935) classification of courtship displays into 'lizard', 'labyrinth-fish' and 'cichlid-fish' types (cf. Chapter XVIII). In the

PLATE IX

page 100

RUFFED GROUSE
Intimidation display.

page 38

CORMORANT DISPLAYING

PLATE X

page 100

BITTERN
Intimidation display.

first type another member of the species is treated as a female, and coition attempted if similar display is not forthcoming; in the second, unless the response is female stimulative behaviour, fighting results. In the third type both sexes display. Leaving aside for the time being the mechanisms by which such displays operate, consider this aspect of the situation: Is it credible that the emotions—such as they are—concerned in these displays can be anything like so clearly defined as the emotions we know as love and hatred, or jealousy? Is it probable, all the facts being considered, that a bird's emotions are so finely poised that it can suddenly switch over from one to another and back again according as the other individual responds in one way or the other? To use a crude analogy, it is as if a truculent fellow going about 'trailing his coat' and spoiling for a fight could in that mood fall in and out of love between fights. It is much more reasonable to assume the presence in the bird of generalised emotional feeling, excitement which becomes tinged with a somewhat more specific tone according to circumstances.

Along such lines another type of bi-valent display can be understood. The 'piping parties' of the oystercatcher will be referred to later, but it may be said here that in spite of much study there is considerable doubt and disagreement as to how they should be interpreted. It is highly probable, in my opinion, that the display produces a pleasant excitement, favourable to breeding, in all the birds which participate, but that it has intimidatory reference for intruders while being sexually stimulating for the pair of mated birds which take part (Huxley and Montague, 1925). Possibly the assiduous gaping and head-wagging performed by fulmar petrels all the time they frequent the breeding cliffs may be explained along such lines. There can be little doubt that general stimulation is appreciated and courted by many birds; the difficulty in discovering what emotion is present in some circumstances is due to the fact that generalised excitement has to reach a certain pitch before it is coloured by a specific emotional tone and shows itself in appropriate behaviour-patterns.

When we review the emotional manifestations which have been mentioned in this chapter we notice, firstly that in a number of species threat and sexual displays are alike and in others they are quite distinct; secondly, that sudden shock or excitement may elicit a totally inappropriate response; thirdly, that some displays, such as the fulmar petrel's gaping, do not seem to be the expression of any single identifiable emotion, and fourthly, that a neutral action may become appropriated to a specific emotion. Neither a Freudian nor a Hingstonian theory can explain these facts. The only hypothesis which is adequate to interpret them is the view that the emotions of birds are not so clearly defined as our own. Köhler (1927) shows that even amongst anthropoid apes sexual behaviour is not so definitely differentiated as amongst men. He writes:

> The sexuality of the chimpanzee is as it were less *sexual* than that of the civilised human being. Often, when two chimpanzees meet one another, they seem to 'sketch', or indicate, movements which can hardly be classed definitely under either the category of joyous and cordial welcome, or that of sexual intimacy.

What is true of apes is still more true of birds. Their emotions are not so specific as human emotions—and this is the main reason why some biologists question whether they have them at all. In man self-consciousness gives a deeper content and more precise reference to the emotions than obtains in any other animal. Perhaps I should explain that I use 'self-consciousness' in the sense in which it is employed by Lloyd Morgan (1930) as applicable to mind at the reflective level.

We have seen that in man, too, confused emotion and inappropriate emotional expression may occur. The phenomenon of flushing man shares with mammals (monkeys) and birds (turkeys, jungle fowl, bateleur eagles). In him it may be caused by such different emotions as joy, anger, love, shame or generalised excitement (Darwin, 1872). The generality of its reference shows it to be akin to some bird displays, but self-consciousness enables human beings to realise the distinction between the emotions of which it is an expression. Hingston uses the flushing of the bare skin areas in support of his postulated 'fear-anger' conflict, but what it indicates is very different—the existence in the organism of ungeneralised emotion or excitement. The more fundamental concept is that of energy; emotion is its psychic colour. If for one reason or another, such as sudden fright, thwarting of activity or conflict of impulses, the appropriate reaction to a stimulus cannot occur, then nervous energy must find expression in some other action with emotional valency.

Early in the breeding season 'confusion of impulses' is apparent amongst birds as different as warblers (Howard, 1929) and penguins (Roberts, 1940a), though even these follow a sequence related to the breeding cycle. It seems that some displays, such as the bowing of terns (Southern, 1938), are just excitatory and work up generalised emotion preparatory to its differentiation in the next phase. As it rises in intensity the concomitant more precise display occurs. In some cases one might say that there is a topographical counterpart to this. For example, amongst herring gulls there is a tendency as time passes for posturing birds to retreat from public, neutral ground to the privacy of the nesting site (Darling, 1938). As intensity of feeling increases more precise localisation meets the emotional demands of the situation. When we consider 'arena' displays the significance of this will be apparent.

A consideration of emotion in human beings supports the view advocated here. MacCurdy (1925) writes:

As excitement increases, the attention of the subject to his own feelings diminishes, action becomes automatic, he is 'beside himself', loses that shade of subjective feeling that ordinarily serves to accelerate, inhibit or direct his thoughts and actions. In short the affect (feeling-tone) goes as excitement increases. Naturally, then, all emotions tend to become alike with excitement, the feeling is degraded to mere inner tension or may be quite absent, as in reckless action. The organism is then just behaving, the guidance of consciousness (self-awareness) is removed. As we shall see later there is evidence for this in psychopathological material: we meet with cases in which no trace of any affect can be discovered or inferred. The patient is simply in tense excitement.

In fact, the patient returns emotionally to a more animal condition. This is the more readily understandable as the emotions are associated with an ancient part of the brain which, according to Sir Charles Sherrington (1940), 'still continues that of man's ancestral and related stock of long ago. It has meant in them, and still does so, fear, rage and passion; it does so also with man'. Injury to it results in torpor. But in man the roof-brain, a hundred times greater in volume, usually is able to exert a large measure of control. If we venture to think of it in picturesque terms we may regard it as where energy is concentrated to be directed or 'boosted' into various channels; in birds more particularly into behaviour-patterns. Sometimes these may be inappropriate, sometimes active response is partly or wholly inhibited.

It has been frequently pointed out that emotion is apt to appear when no adequate response to a situation is possible. Flight from danger, for example, decreases fear, and I have noticed that timorous feelings during an air-raid disappear when one is actively employed in some necessary work. Rock climbers who experienced no fear on a difficult 'pitch' sometimes feel it afterwards. Hesitation increases emotion, but 'I hadn't time to be frightened' is a common phrase. In short, emotion appears when instinct has no immediate and adequate outlet, when the internal state of the organism is out of harmony with the behavioural field, or there is conflict within the personality

(Angier, 1927; Dembo, 1931). It may arise through internal conditions, such as the state of the endocrine system, causing drives and creating needs, or as the result of impediments to action in the environment. For example, chiffchaffs and willow warblers display and copulate in autumn (Brock, 1910; Palmgren, 1934; Lack, 1939 a) and the autumnal 'dances' of stone curlew (Selous, 1901) and various waders (Lack, 1941 b) are apparently due to the migratory impulse not yet finding full expression in passage flight. Howard (1929) holds that bodily movements in posturing are in inverse ratio to the intensity of feeling. It is true that a male bird, unable to join in the sexual pursuit of the female because of her sluggishness, quivers his wings just as a caged cuckoo in autumn quivers his, and that when a male's sexual curve is not sufficiently high to enable him to react fully to the female's invitation he agitates his wings, but male birds may join in sexual flight and yet be unable to copulate. While there is much truth in Howard's axiom the relationship between feeling and its expression is more subtle than it would suggest. Posturing acts as a valve for emotion, sometimes enabling it to increase, at other times permitting it to find release. Sometimes, indeed, the display seems to augment, or even to cause, the emotion as when pouter pigeons inflated by mischievous children strut about—if, indeed, we may presume that emotion is aroused in such circumstances. But once again we notice how naturally descriptions of emotion call forth analogies involving the concept 'energy'. Energy and mind must be regarded as ultimate and incommensurable (Sherrington, 1940), but the fundamental importance of the emotions and also an explanation of the obscurity of their nature may be found in the fact that here mind and energy are interwoven and co-operate.

If 'injury-feigning' illustrates what happens when there is conflict within the organism, courtship posturing shows us the normal operation of the emotions in regulating birds' energies into appropriate channels and achieving synchronisation between the sexes. The emotional processes of birds during courtship may be compared with the situation which arises when two people have a heated argument. As passion mounts they 'forget themselves' and eventually come to blows. The words and gestures which constitute the 'display' of the parties to the altercation express their emotions, and each by his 'display' arouses further increments of emotion in the other, accompanied by the secretion of adrenaline and other physiological changes (Cannon, 1929), until at length passion explodes into violent action. A person, picking a quarrel, may 'work up' his own feelings to a certain point, and apparently so can birds. When a female bird is the object of display we may believe that if she is in the requisite physiological condition her feelings mount as do those of a person with whom a quarrel has been picked. In mutual courtship and connubial display feeling is maintained at a high, and in this case desirable pitch, for a prolonged period.

CHAPTER IX

GREETING AND OTHER CEREMONIES

Recognition amongst birds—The respective functions of morphological characters and behaviour—Invitatory ceremonies—The importance of appeasing and greeting ceremonies—Nest-relief ceremonial—The emotional basis of transition rites—Crude and elaborate ceremonies—Nest-relief amongst fish.

IN the first chapter we noticed that when a gannet returns to its mate the birds engage in 'billing' or 'bowing'—often both. These ceremonies have their origin, no doubt, in the sex relationship of the birds, but as we see them performed they do not appear to have an immediate or direct relationship to that function. In so far as they are performed when male and female meet we may consider them to be 'greeting ceremonies' or even 'appeasing ceremonies', but they may also be regarded as connubial posturings, probably fostering and maintaining the synchronisation of the two organisms. If so, we have here a warning of the difficulties besetting any classification of bird ritual. We may designate a series of display-types, but while certain examples will seem to have a particular well-defined function others will be found to have no such clear-cut functions or emotional import. Some greeting ceremonies, as Lorenz (1935) shows, have an independent function as recognition and reconciliation signals, but others shade into and can scarcely be distinguished from courtship and connubial ceremonies.

The psychological problem involved in birds' recognition of one another is complicated and, as yet, inadequately explored. Song sparrows (Nice, 1937), black-headed gulls (Kirkman, 1937) and crimson-crowned bishop birds (Lack, 1935) recognise near neighbours but not strangers, whereas the individuals of a jackdaw colony are all able to recognise each other individually (Lorenz, 1931, 1935). But as to how birds wanting mates recognise the sex of another bird there is a good deal of obscurity. The female red-necked phalarope at first reacts with 'wing-rattling', ceremonial flight and characteristic call either to males or females of her own species or even approaching ringed plover, Lapland longspurs, or purple sandpipers (Tinbergen, 1935). It seems that stimuli which operate at a distance are comparatively generalised and that when the phalarope is closer to the intruder discrimination is facilitated and the behaviour correspondingly modified. When Noble (1936) attached the 'moustache' characteristic of the male flicker to a female her mate chased her as if she were an intruding male, thus showing how important are morphological characters for sex identification, but Tinbergen (1939a) points out that Allen's (1934) conclusion, based on a study of the ruffed grouse, that birds are not sex conscious is erroneous, as this bird recognises sex by behaviour. This is also the case in pigeons (Whitman, 1919), song sparrows (Nice, 1937) and penguins (Roberts, 1940a).

In many species in which there is no pronounced sex-dimorphism, behaviour undoubtedly plays an important rôle in recognition. Bennett (1939) conducted a series of experiments on ring doves—which recognise each other individually in captive flocks—and discovered by altering the coloration and contour of the plumage that recognition seemed to be based on behaviour. Such evidence emphasises the importance of posturing and the actual movements involved in display, and suggests that it is by their deportment that such birds as robins recognise their mates at distances of thirty yards or more (Lack, 1939a). Only by studying the individual species is it possible to determine the mechanism by which recognition is achieved and whether it is mainly morphological, behavioural or both.

GREETING AND OTHER CEREMONIES 109

Recognition is obviously of most importance to birds in connexion with the admission of other birds to the nesting territory or the nest itself. Not only must prospective mates or rivals be identified as such but completely different modes of behaviour, enticement or repulsion have to be brought into operation as soon as possible. Thanks to Tinbergen's work (1939 *a*) the procedure of snow buntings is well known. It will serve as an example showing how an invitation ceremony follows on recognition.

A cock snow bunting in his nesting territory crouches in a threatening attitude and utters a call *peee* at the approach of another snow bunting, whether near or far, male or female. He may remain thus for a considerable time and often until the other bird is only a few yards away. If it declares itself a rival by threatening behaviour, then sparring begins, but if the intruder loiters and is identified as a female an invitatory pose is adopted. The male runs repeatedly from her with his piebald wings and tail spread, holding himself in an upright position. Frequently the female is not really frightened by the male's initial threat-action, but alights quite close at hand. There she preens and picks amongst the herbage in the intermittent and incomplete way indicative of substitute activity and inhibited emotion. Soon they are foraging together and the male may try to copulate.

This procedure may be compared with the ritual by means of which the pairing-up of various species of heron is achieved. According to Noble, Wurm and Schmidt (1938) and Allen and Mangels (1940), American black-crowned night herons return to their arboreal nesting colonies early in spring and each male bird selects a territory. Before much time has elapsed it will be seen that these cocks are behaving in a queer way; they break twigs off the branches and loudly snap their mandibles, still holding these sticks, at the same time moving their heads rhythmically up and down. Alternatively they may stand alone on a nest platform or potter about amongst the branches while performing thus. The twigs patently have a ceremonial function, for although they may sometimes be laid on the nest platform they are usually dropped. It is not to be imagined that these plucked twigs are intended as a gentle hint to the female that it is time she interested herself in the domesticities, though twig-presentation in some species may have a sexually stimulating effect. Such anthropomorphic conceptions are misleading. The manipulation of twigs on the part of the black-crowned night heron is another instance of the tendency to fidget with materials put to a good use. From a nervous habit it has evolved into a meaningful ceremony.

Other birds, or the same birds at other times, perform another ceremony. Watch a black-crowned night heron moving about a tree and we presently see him take two or three steps forward, halt, hunch up his back and lower his bill nearly to the level of his pinkish feet; then, while raising one of these he emits a snapping or clicking sound from his throat, immediately followed by a prolonged hiss. He repeats this performance eight or ten times per minute, sometimes raising the same foot, sometimes each foot alternately.

We shall have occasion to return to a consideration of this American heron's ceremonies later, but now it will suffice to notice the effectiveness of this symbolic code. The female is beguiled by the male's ceremonial into overcoming her repugnance to trespassing on another bird's preserve and in many cases she enters and mates with him. The male's gesticulations have a remarkable power over her. It would be a mistake to call it mesmeric, but there is a subtle element about it which approximates to compulsiveness. It is as if in the female's mind there is an innate recognition of the signal and acquiescence to its meaning which almost necessarily entails response. The male's posturing in these ceremonies is not merely interesting or attractive to the female, but seductive, just as the female cormorant's gaping is to the male. The female's mind is apparently innately attuned to the male's display—hence its effectiveness. Nevertheless, she may pass by a gesticulating male; it seems possible that this happens when the sexual cycles of the two birds are not sufficiently synchronised.

The courtship display of the European night heron broadly resembles that of the American species (Lorenz, 1938). The male chooses a site, begins to build, and perched on the spot where the nest will be, calls for a mate. This he does by treading a courting dance, first on one foot, then on the other, with a peculiar weaving action. He lowers his head occasionally and utters his courting cry—a sound like steam escaping. Whether another bird be in sight or not the performance goes on, but the intensity of it all increases greatly when another night heron approaches, be it male or female, for this species distinguishes the sexes not by plumage but by behaviour. Only by degrees does the female qualify for admission to the territory. On the one hand she has to overcome her reluctance to trespass; on the other the male has to learn to inhibit his territory-defending reaction. The distance at which this is aroused gradually decreases, and it seems that the mating cycles of the two birds approach synchronisation during this period. When striking distance is reached a special appeasing ceremony is used. The male lowers his beak and stretches forward his neck, uttering a low disyllabic call. In this attitude the raised black crest with the three white plumes is presented most impressively to the other bird and she reacts with the same display in token that her repelling impulse has been suppressed. The crest is not displayed for any other purpose according to Lorenz, but Noble, Wurm and Schmidt deny this in regard to the American race. They state that the plumes have no appeasing functions but are used in sexual or epigamic display, and believe that the female responds because she is stimulated by the plume erection. The important part played by vocal sounds in ceremonial of this type was shown by plugging the ears of the black-crowned night heron, thereby preventing synchronisation of the greeting ceremonies. Allen and Mangels (1940) found that there was no significant difference between night and day behaviour during the courtship period. This suggests that these writers attach undue importance to some of the visual stimulatory features, such as the pink legs of the herons.

Invitatory ceremonial in which the male bird stretches up his neck and utters a strange note has been described in regard to the Louisiana heron (Huxley, 1924), the green heron (Townsend (1928), and the common heron (Verwey, 1929, 1930). This bird may stretch head and neck vertically upwards, then bring them down, still pointing upwards, while he bends his legs and lowers his body. He calls *hoo* at the maximum extension upwards and gurgles *oooo* as the head comes down. Or he may lower his head towards his feet, snapping the mandibles together. The note ceases when the female approaches, stretching is repeated a few times, and as she draws near the snapping ceremony is reproduced up to forty times. A male may call for a female week after week and then drive her away with an unrepressed defence reaction unless her advance is very slow and cautious. Only gradually is territorial defence subordinated to the sex urge so that the female is admitted; even then 'misunderstandings' may ensue and tiffs and 'bill-sparrings' take place. Presently, however, the birds caress and nibble each others' bills in what Lorenz calls, in the case of night herons, a 'reconciliation ceremony'. Eventually he presents a stick to her ceremonially, both birds having crests and plumes raised. Sometimes the female now performs the 'stretching' gesture. She builds the stick into the nest, and as the male sees building proceed he becomes greatly excited and fondles her feathers with his bill. She responds and coition takes place. Stick presentation in the heron has analogies with the courtship feeding of the robin, for although copulation and feeding are separate they occur within a day of each other. Stick ceremonial may also be enacted when the birds have been together on the nest for some time and sometimes when one bird relieves the other on the eggs (Huxley, 1924). In like manner the 'appeasing ceremony' of the European night-heron, used first on pairing up to inhibit the other's stabbing reaction, is performed as a greeting regularly during the breeding period whenever one bird returns to the nest.

The important function which the appeasing ceremony may serve in the life of birds is illustrated by the behaviour of a pair of storks which mated at the Schönbrunn Zoo. The female was a white stork, the male of the black species. The latter bird does not often 'klapper' but uses a side-to-side and up-and-down neck motion with a peculiar whispering as a greeting ceremony at the nest. Although the birds had been paired for years there was continual misunderstanding between them owing to the difference between these ceremonies. When rock doves mate with wood pigeons there is similar confusion (Lorenz, 1935).

Appeasing ceremonies or postures may have relevance rather to the position of birds in the social hierarchy than to mating, although undoubtedly the relative rank of the two birds concerned is important during and after courtship. A jackdaw will turn the back of its head to another member of the flock in token of submissiveness (Lorenz, 1938). One of four Forster's milvagoes in the London Zoo has been seen lying on its breast before another, exposing the back of its neck to it (Fisher, 1939).

The greeting rites of storks attracted attention in remote times. Ovid (*Met.* VI, 97) and Philostratus (*Ep. ad Epictet.*)

mention the 'klappering' of the birds and Petronius (*Satyricon*, LV) refers to

> The stork, welcome wandering guest
> With slender legs and bill rattling like castanets.

Dante (*Inferno*, XXXII) compares the chattering of traitors' teeth in the icy Ninth Circle of Hell with the clattering of the stork's mandibles:

> mettendo i denti in nota di cicogna.

Klappering, as I have seen it performed by birds in Holland and Yugo-Slavia, varies from a mild rattling of the mandibles to an elaborate performance in which the bird bends its head back until the occiput all but touches the back—an attitude similar to that of the courting cormorant. It then brings its head forward and down, vigorously rattling its beak the while. The mate responds in somewhat the same fashion. Alarm, anger, intimidation or pleasure may be expressed by klappering, but the sequel makes the meaning clear. A tame stork which greets all intruders with this instrumental music subsequently shows its welcome of a friend by picking up some nesting material (Heinroth, 1924–33).

The ceremony of nest-relief conducted by the grey heron is somewhat similar to the snapping or appeasing 'figure' of pairing-up. With a swish of pinions and vigorous flapping the male alights on the rim of the nest. The incubating bird, uttering a medley of clamorous cries rises and stretches her neck upwards. Now they stand beak to beak, both emitting hoarse shouts. Then the male bends his head, with the crest raised, snaps his mandibles several times and settles on the nest. The female presently departs. Variations in the ritual occur. When the male stretches his neck upwards, looking almost like a figure seen in a distorting mirror, he may then raise his crest and

PLATE XI

page 114

HERONS

Greeting ceremony.

PLATE XII

page 119

BROWN PELICANS
Nest–relief ceremony.

page 118

WHOOPER SWANS
Mutual greeting ceremony.

agitate his wings. The female acts in much the same way, but her manner is usually more subdued than his (Selous, 1905 a; Huxley, 1924).

The greeting or appeasing ceremony, with its associated call, may· also play a considerable part in the relationship between parents and young. The discernment of movement ante-dates the recognition of form so that it is natural that posturing which is for the young a signal to prepare for dinner should be evolved. There is fierce competition amongst young herons and the weakest tend to be neglected. The nestling which perceives the approach and arrival of the old bird and importunes it first and most vigorously has the best chance of survival—other things being equal. Lorenz has shown that young herons will actually attack the parent if he accidentally omits the appeasing ceremony. In some cases the greeting gestures of young birds may be designed to enable the parents to find them more readily. Nightjar chicks merge into their surroundings and as they grow tend to move away from where they were hatched. They have a special ceremony during which they stretch their necks upwards with head and body horizontal, at the same time raising and spreading their wings frequently while uttering a murmuring note (Heinroth, 1909).

Writing of the gentoo penguin Roberts (1940 a) shows the importance of appeasing ceremonial:

If a penguin occupying a nest is approached by its mate, it will tend to react aggressively in just the same way as it does towards a strange bird. Indeed, it is not uncommon to see one member of a pair peck at the other if the relieving bird arrives too suddenly at the nest because it has been attacked while passing close to other birds. The elaborate bowing and hissing of both sexes which accompanies the change over in incubation evidently prevents the birds belonging to the same nest from fighting each other. It is possibly quite distinct in function from the mutual bowing which occurs before the eggs are laid. Unfortunately, I did not appreciate this possibility while in the Antarctic, and my notes do not disclose any significant differences between the two ceremonies.

This may be compared with Ponting's (1921) description of Adélie penguins:

> The formality of relieving guard was accompanied by a good deal of ceremony. The returning cock, spotlessly clean and glossy from immersion in the sea, would, on reaching his nest, bow several times to his partner—making a soft, gurgling sound in his throat as he did so. He would describe circles in the air with his head, still gurgling; and the hen would raise her head, stretch her neck, and describe circles and gurgle too. This would continue for some time.

He describes the female as invariably needing much persuasion to induce her to leave the eggs. Ponting's identification of the sexes was guesswork, but as male and female perform alike this does not alter the value of these observations.

In an earlier chapter we considered instances of ceremonial gaping in courtship and connubial displays. This gesture is also used in greeting rites. For example, when a hawfinch returns to the nest its mate salutes it with open bill and quivering wings (Bird, 1935). Possibly, however, this represents food solicitation, but even so it has a ceremonial character. The greeting ceremony of red oven birds is most elaborate. When a pair of these rejoin one another after separation, the home bird, seeing its mate appear, starts emitting loud and regular notes, or sometimes a measured trill; but when the other joins in with single notes the first bird changes its tempo and utters lively triplets strongly accented on the first note. As they sing they keep time, each to its own music, the one fluttering its wings rapidly to accord with its quick succession of notes, the other flapping a more deliberate rhythm to its measured song. The welcome concludes with three or four clear, ascending notes (Hudson, 1892). The precocious young begin their musical education early, for they 'are constantly heard in the nest or oven apparently practising these duets'.

A common screamer approaching its sitting mate starts calling when a few yards away. The other stands up and answers, and both call together for about fifteen seconds, the male in a deeper tone than the female. There is mutual preening and each in turn examines the eggs. The vacating bird then spends two or three minutes jerking nesting material over its shoulder in the manner which we have seen to be characteristic of many birds when leaving the nest (Stonor, 1939).

Greeting and courtship—or connubial—antics are closely intermingled in some displays, as we have already noted in the case of the gannet. We have space here only for two examples. Great skuas will occupy stances some distance apart, and every now and then one will visit the other. When a skua makes a call it bows its head solemnly to the ground while the visitee stands facing it with bill elevated; then the visitor turns round, and raising its wings over its back makes off in a series of hops. This procedure may be seen three or four times in the space of an hour (Selous, 1901). Here it is difficult to discriminate between the greeting and courtship elements. According to Southern and Venables (1936) the elevation of the white patches on the wings is principally a greeting signal, but is also used in the early courtship period. The wing-raising ceremony of the purple sandpiper is employed as a greeting and on other occasions. Keith (1938) noticed it given by mistake to a snow bunting.

Another skua ceremony is thus described by Selous:

The two birds have been circle-soaring one above the other, and are now at a considerable height above one of their chosen standing-places, when the lower one floats with the wings extended, but raised very considerably— half-way is uncommon. As it does this it utters a note like 'a-er, a-er, a-er' (a as in 'as'), upon which, as at a signal, the other one floats in the same manner, and both now descend thus, together, to the ground. Standing, then, the one behind the other, at about a yard's distance and faced the same way, both of them throw up their heads, raise their wings above their backs, pointing them backwards, and stand thus for some seconds fixed and motionless, looking just like a heraldic device. At the same time they utter a cry which sounds like 'skirrr' or 'skeerrr'. The foremost bird then flies off, and is instantly followed by the other.

In this ceremony one bird of the pair acts like a replica of the other. Whooper swans, too, perform identical posturing as greeting ritual when the cob returns to the nest, the pen to the cygnets or at any other time when their feelings are aroused. A transcript from life is better than many generalisations where bird behaviour is concerned, so here are my notes of the ceremonial, already published elsewhere (1940):

Suddenly, for no apparent reason, there is a commotion and one bird of a pair starts in pursuit of another and they flog the water with thrashing wings and splattering feet. The partner of the angry bird flaps after him. He desists from the chase, and now the two stately birds swim face to face with tail feathers erected. A wild clangour rings over the lake as both swans, with beaks raised and white wings uplifted co-operate in a resonant duet, the pen replying to the quick 'honking' of the cob with repeated notes half a tone lower. There is I know not what of wild and sweet in the song of the 'swans with their exultant throats', so different from the plaintive imaginings of the poets. They flap their wings and repeatedly dip their necks in sinuous simultaneity, then both stand up in the water raising their bodies high—and I am reminded of the 'penguin dance' of grebe and diver and the 'wing flappings' of ducks and guillemots when sexually excited. Both now swim together with necks and heads keeping time in a rhythmic forward motion. All this time the clamour resounds over the water and away to the ragged hills, but as the activities gradually subside the trumpet-notes lose their rapidity and excited urgency and soon the birds are swimming placidly some little distance apart, the male uttering occasional calls which are immediately answered by the female.

The swans frequently stooped their necks as if to drink, so that the movement had the appearance of being due to nervous tension rather than a desire to quench thirst. Once a bird swam for an appreciable time with wings half-spread on the water and head dipped, looking crippled. It was as if it were doing the 'broken-wing trick'. Christoleit (1926) describes this posture as being adopted when two pairs try to intimidate each other. There is frequently the most amazing mutuality in these ceremonies (Selous, 1913–16); at times it is as if a single bird were reflected in a mirror, at other times the 'reflexion' follows with a short time-lag. Such notes as these are frequent: 'Both "crow" simultaneously, both then bob their heads, both move heads and necks forward stiffly and regularly "crowing" together, both put their heads into the water as if drinking, both bow.' The effect is very strange.

The mutual salutations of whooper swans may be called greeting ceremonies as they are characteristic of reunions, but they are so frequently used that they must be considered much more than appeasing or recognitional posturings. Swans mate for life, but as with other pairs of birds which are perennially faithful to one another display is indulged in each year. So far as whooper swans are concerned these connubialities occur not only when a bird returns to its mate in the ordinary course of events, but also when some incident occurs to excite them. Thus whenever one of the pair I watched began to chase an intruder the other flapped off to its support with great webbed feet striding on the water, and as the offender withdrew the swans would swim to each other as I have described, salute ceremoniously and appear to exchange mutual felicitations on their achievement. A similar ebullition of emotion is observable when a third bird approaches a mated pair of red-throated divers.

We have seen that recognition, greeting, nest-relief and other ceremonies are often not readily distinguishable from one another; or, rather, one ceremony may have these and other 'meanings'. Are not several meanings inherent in the ceremony of the Changing of the Guard? Instances already mentioned show that recognition and appeasement enter into nest-relief ritual just as they are in a sense involved when we shake hands.

Brown pelicans perform a nest-relief ceremony worthy of special notice, but it is believed to occur only when the male bird takes over from his mate (Chapman, 1908a). Returning from fishing, resting or bathing the pelican flies in with slow, majestic strokes and alights close to the nest; then, pointing his great bill to the sky and presenting a very odd appearance, he advances slowly, waving his uplifted head and bill like a banner from side to side. At the same time the sitting bird pokes her beak vertically into the nest, twitches her half-opened wings and utters a guttural *chuck*. This is one of the very few occasions when an adult brown pelican makes any vocal utterance. The advancing bird brandishes his bill on high, back and forth five or six times, and pauses. Then—a dreadful anticlimax—both birds begin to false-preen like mischievous schoolboys caught at some embarrassing moment making pretence of being busy with something other than the real matter in hand. A moment later the incubating bird rises and vacates the nest, and the new-comer settles into her place.

Nest-relief as performed by the buff-backed heron is a simpler affair. The returning bird climbs down to the nest and the other extends her head and neck. (I am assuming the sex.) He lays his neck across and along his partner's, and so they remain for a few seconds while a crooning sound is uttered by one or both. Gradually the incubating bird gives way and her mate takes her place while she departs through the branches (Farren, 1914). When yellow-crowned night herons change places they erect plumes (Nice, 1929) and Louisiana herons also elevate their plumes conspicuously, the relieved bird presenting its mate with one or more sticks (Huxley, 1924). Black-crowned night herons accompany plume erection with bill-rattling and a soft *wok, wok* call. Both members of the pair act alike (Allen and Mangels, 1940).

Renunciation of the eggs and the never-to-be-too-much-sat-upon mood in regard to them often involves an emotional wrench, and sometimes the sitting bird is very reluctant to vacate the nest. While some birds have evolved ceremonies which enable the transition to be made without misunder-

standing or fuss, others, to their disadvantage, have no such
etiquette. What could be more decent and orderly than the
woodcocks' polite ritual enacted on the brown, leafy carpet of
the coppice? When one bird returns to the nest the other walks
up to it and they lay their long bills slantwise across each other
like two swordsmen about to engage. Thus they stand for nearly
a minute. Then, the solemn rite having been duly honoured,
the newcomer takes his (or her) place on the eggs while the
other goes off to feed. The ceremony, it should be noted, like
many other such, takes place at dusk (Seigne and Keith, 1936).[1]

It is not by accident that there is a resemblance between some
of these rites and the Changing of the Guard. Admittedly there
are innumerable differences of detail between the military ceremonial in Whitehall and the etiquette of the woodcock in the
dusky spinney, but both are the manifestation of the emotion
which concentrates itself about what anthropologists call transition rites or *rites de passage*. Changes in status, such as sexual
maturity or marriage, require to be regulated by ceremonial.
The idea underlying all discipline—and what is ceremonial but
the discipline of emotion into ordered and socially harmless
activities?—is that reactions, whether of a man, a tribe or a
nation, should be so drilled by habits of will and action that in
times of emotional stress control is maintained. Ceremony is,
indeed, man's antidote to disorder and frenzy; and not man's

[1] The female nightjar broods during the daylight hours and is relieved by
the male in the evening, according to the observations of Selous (1905 a) and
Lack (1932); but Selous decided that the female heron starts sitting between
4 and 5 p.m. and continues all night until between 6 and 9 a.m. (1905 a).
Wilson's petrels incubate for approximately 48 hours at a stretch and always
change places at night (Roberts, 1940 b). The female song sparrow comes off
the nest to feed every 20–30 minutes (Nice, 1937), but the eider duck vacates
her eggs for only a few short intervals during her 28 days' incubation, and
it is said that the argus pheasant remains on her eggs for the 24 days of the
brooding period (Beebe, 1922). Laysan albatrosses relieve each other after
about a fortnight but sometimes sit for 30 days (Hadden, 1941).

only, for it exercises exactly the same function amongst other creatures. We shall see later how much this is true of a specific form of the ordering of emotion—the dance.

In those species which have no definite change-over ceremony or for one reason or another omit it, clumsiness and misunderstanding may occur. A nightjar will push his mate from the eggs, but it is done in such an orderly way that it almost constitutes a ceremony. The male flies in uttering a curious and significant note, *quaw-ee, quaw-ee*, and the female churrs softly in response. He settles by her side, 'snoozling against her', and they sit churring together, swaying their bodies from side to side. Suddenly the female departs leaving the male on the eggs (Selous, 1905a). Amongst avocets the sitting bird usually goes to meet the approaching mate, both perform bowing motions and utter twittering sounds; then the bowing changes into throwing straws and shells about and the one takes the other's place (Makkink, 1936). But when the partner persists in incubating the other pushes with his cobbler's awl bill, calls vehemently, flicks pebbles as big as marbles at her as he parades around and eventually sits down and thrusts her off (Booth, 1913). Black-headed gulls will also try force when anxious to brood, though, as has already been mentioned, the impulse often is diverted into collecting material for the nest. A returning bird, thwarted of its desire to sit on the eggs, was seen to go in search of material and eventually brought back a weed which it dropped on the edge of the nest. Then it tried to push the recalcitrant one off with its foot. It gave the crooning call, used to summon the hen or chicks to food, but this, too, failed. Off it went collecting again. It reappeared with a short stalk in its bill, found the nest had been vacated, and started to incubate (Kirkman, 1937). The black-headed gull illustrates the disadvantages of being without a definite code. This bird may act in any one of six ways—including absence of utterance or display—on returning to the nest, and its mate at the nest may respond with any one of these, not necessarily that adopted by the other. It is not surprising that confusion ensues.

Herring gulls have these matters rather better organised than black-headed gulls, for they possess a nest-relief ceremony, but like the ritual of the grey heron the same ceremony, or one closely resembling it, may be used on other occasions. The procedure is, in fact, similar to that which initiates the nest. Early in the season male and female squat on the ground with bodies tilted and make as if they were gobbling up food from the ground. Their actions wear the soil into a hollow in which the eggs are laid in due course. This rite is sometimes performed on neutral ground, that is, away from appropriated territory, and then as many as eight birds may participate (Darling, 1938). The inference is that the movements evolved from coitional adjustments became the initiation of nest-building and are now, and perhaps increasingly becoming, a nest-relief ceremony. Thus these and many other ceremonies originated from and evolved out of elementary, primordial movements connected with reproduction and nutrition. When we later consider the origin of communal ceremonies the infectiousness of such rites as these will be seen to be a phenomenon of prime importance.

As our study of the gannet revealed, in emotional situations or times of nervous tension there is a tendency for birds to toy with pebbles, leaves or grasses. Often preening occurs. As facts quoted in subsequent pages show, birds which perform on 'courts' usually clean them of every leaf and stem with extravagant care, and even in courtship ceremonies there are traces of this nervous habit. The cock grey peacock pheasant prefaces his display by scratching the ground for a tit-bit; having found one he holds it in his bill and attracts the hen by making a series of clucking sounds. As she draws near he sinks his breast and expands tail and wings. When she is only a foot or two away he

jerks the grain of seed or whatever he is holding towards her
and deploys the gorgeous tail and wings to their fullest extent,
at the same time pressing his head sideways against his wing
(Seth-Smith, 1914). The peacock scratches with his feet during
his display. But when nest-relief occurs fidgeting with odds and
ends of material lying close at hand is a relatively common
feature. There is space to add only a few instances to those given
in Chapter II. Coots and dabchicks bring weed to the sitting
bird when relieving her on the nest. On leaving the eggs a
male turnstone walks slowly away, all the time picking up and
throwing over his shoulder small pebbles in the direction of the
nest. The female then approaches, acting in exactly the same
way (Gordon, 1936). An elaborate ritual is sometimes performed by stone curlews. During the last four or five days of
incubation the birds take turns of duty on the eggs. After sitting
for a time the cock calls softly and his mate comes running to
him over the heath. He rises slowly, and, picking up a small
pebble, offers it to her with a bow; then walks off in a slow,
dignified way to his observation-post some yards from the nest.
After the hen has been sitting for over an hour the cock calls
and she answers. When he returns the stone-presentation ceremony is re-enacted, the female picking up the pebble as the male
draws near. If he does not take it in his bill she lays it down
between the two eggs, rises, bows and hurries in a crouching
posture from the nest (Hosking and Newberry, 1940). Among
white or masked boobies the first manifestation of breeding
activity is when one picks up a small stone and places it on the
ground in front of the other. It accepts the offering and slowly
and deliberately places it on the ground again. This procedure
may continue for an hour or two. The ceremonial of the blue-
footed booby is exactly similar (Murphy, 1936). Somewhat
comparable behaviour is reported of gannets; besides the bowing and fencing which take place when the birds meet, festoons

of seaweed may be passed from beak to beak before the mate takes over the domestic duties. Perhaps the prominent part invitatory, greeting and nest-relief ceremonies play amongst herons, gannets and boobies may be correlated with the development of the bill as a potentially powerful weapon and the consequent importance of conciliatory gestures.

Nest-relief ceremonies are also performed by fish. Species of the Chromide group swim up to their mates on guard at the nest using peculiar movements which have the effect of stimulating the sentinel to vacate his (or her) post and make way for the other. Such ceremonies, whether they occur in birds or fish, are much more than polite salutations. There is good reason to believe that the sight of the appropriate antics and postures acts like a switch, facilitating the change-over by countering the broodiness of the incubating bird or the 'dutifulness' of the guardian fish.

This discussion shows that a bird employs ceremonial, not only to express its emotions but to influence the emotions and control the behaviour of other birds. Without the ceremonial code there would be misunderstanding and confusion as between male and female birds, and the sexual cycle would be thwarted and its pattern broken up. What we may regard as an elaborate system of ceremonial signalling has been evolved in order that the rhythms of male and female might be synchronised and the perpetuation of the race thus assured. It is a thing to be thankful for, that Nature, having strictly practical ends in view, has achieved the creation of a wealth of beauty in carrying them out.

CHAPTER X

MUTUAL AND RECIPROCAL CEREMONIES

Examples of mutual ceremonial—The aquatic displays of great crested grebe and red-throated diver—Mutual rites of swans and penguins—Aerial duets —Vocal duets—Extreme reciprocity—The function of imitation in bird display—The achievement of synchronisation—Tension the mainspring of progress.

IN earlier chapters we commented upon the devices used by birds to maintain and heighten each other's emotional tone, especially commenting on the bestowal of gifts and the exhibition of a lurid mouth as sexual stimulants. In the preceding chapter our survey of greeting ceremonies showed the importance, during breeding, of male and female being *en rapport* with one another. It is becoming ever clearer as we review the ceremonies of birds that the successful raising of a brood depends upon the happy mutual adjustment of the sexes psychologically and especially emotionally. A bird, no less than a man, is restricted in its activities, not only by its environment and its structural adaptation to that environment, but by what we may call a psychical framework, due partly to the exigencies of its own organism, and also, in part, to the individual's accommodation of himself or herself to social life— particularly the relationship to the other sex.

It is but a small step from greeting and reciprocal ceremonies to 'mutual display'—the term used when male and female together execute similar, or approximately similar, ceremonies together. We have already noticed that our 'type specimen', the gannet, is one of the species exhibiting mutual display. Performances of this kind are usually characteristic of species in which the sexes resemble each other, but by no means all species in which male and female are similarly adorned have mutual display. Penguins, though so alike in outward appearance as to be indistinguishable, have unilateral as well as mutual ceremonies, and we have already noticed how the behaviour of the two sexes differs in the courtship of such species as rooks and terns. The one generalisation which we may make is that identical posturing is very rare or unknown in those species in which the sexes differ considerably in appearance. We can hardly consider as an exception the rare instances in which male and female warblers have been observed to perform simultaneously (Howard, 1907–14).

Before going further with our study of mutual display let us consider a few examples.

After chasing each other in wide circles, so that the air rings with their loud calls, cock and hen cedar waxwing perch together on a branch; then each hops to the side and back again, and every time they come together they lovingly touch bills. But often the dance is more elaborate. The cock flies off and presently returns with a bright berry in his bill. Perching beside his lady he passes the berry; she returns it, and so it goes back and forth between them a dozen or more times in succession. Let it not be thought that the berry is just a gift 'from the bridegroom to the bride'; eventually it is returned to him, and he either drops it to the ground or eats it (Crouch, 1936). Thus it is not used primarily as a comestible but as a symbol and a means of attaining and expressing synchronism of the organic rhythm.

A most remarkable rite will be considered in detail later—the courtship of Gould's manakin, a bird of the Central American jungles. The male attracts the female to a 'court' which he has previously prepared, and when she arrives they join in a dance together. She collaborates with him in springing back and forth across this cleaned space and thus indicates that he is the partner of her choice—albeit perhaps a temporary one. Then they go off into the forest together.

The great crested grebe and red-throated diver enact strange mutual ceremonies. In several respects the posturings of the two species show common features, as, for example, in what has been styled the 'penguin dance'—an ecstatic race on the surface of the water in an upright or nearly upright position. In the case of grebes the performance follows some such lines as these: Male and female approach and touch with their beaks, then the female dives and brings up a morsel of weed, which she soon drops. The male now goes under and appears with a larger piece and they face each other thus. Suddenly both leap upright in the water looking like two penguins fronting each other, with their beautiful, silvery white bellies exposed. In some instances the female seizes the weed and, both holding ends, they walk on the surface of the water this way and that for a little while (Selous, 1914). The details of the performance may vary a good deal (Huxley, 1914, 1923); for instance, the penguin appearance may be prefaced by a dive and be not so much sudden as with the effect of 'growing out' of the water. A similar kind of antic is executed by the Slavonian or horned grebe (Selous, 1901–2).

When red-throated divers are sexually excited they may rear up suddenly and run swiftly on the water in an upright position, one after the other (Selous, 1927), having 'a constrained expression or appearance as though they were discharging a duty rather painful than otherwise, but which they owed to themselves and society' (Selous, 1914); or they may race side by side like bottles on end, dive and repeat the performance in the opposite direction. This side-by-side penguin racing may be carried out when the pair are alone on a loch, but it also occurs when another bird is present. The arrival of a third party may occasion the follow-my-leader display.[1] The emotions of a pair evidently tend to be stimulated by the mere proximity of a third, quite apart from any question of territorial rivalry—another fact which throws light on the evolution of communal displays.

'Penguin racing' occurs as a nest-relief ceremony (Selous, 1927), and the upright attitude is also assumed when one bird chases a rival, though the birds could move much more rapidly and comfortably submerged or in the air. Selous regarded the posture as being the true nuptial display as it makes the most of the birds' striking underparts. In many other species, such as the ducks, the exhibition of this area is a characteristic epigamic gesture, used in sexual display rather than rivalry. An attitude which we might call 'standing on the water' approximating to the 'penguin' position is adopted by whooper swans, although unlike ducks and grebes they have no striking breast and belly patterns to show off. Sexual selection, on Selous's view, has brought about the frontal attitude, and now it is not necessarily sexual excitement which stimulates the birds to adopt it. Certainly various kinds of excitement may make a bird assume a strained attitude or perform awkward manœuvres. Black guillemots will chase each other beneath the surface, or in the air, but also, not infrequently, they scuffle together on the surface, half in and half out of the water, as common guillemots sometimes flap and bump along the wavelets when suddenly surprised. It has seemed to me while watching them that they act as human beings do under intense excitement or the stress of strong emotion—doing some action in a difficult or clumsy way because they are too agitated to control, or realise the absurdity

---

[1] A family of great northern divers has also been seen running back and forth in follow-my-leader order over a quarter of a mile stretch of water (Hatch, 1892). Adélie penguins perform swimming games in long files (Ponting, 1921).

of, their actions. The affinity with 'injury-feigning' is obvious. I have noticed a black guillemot, neither chased nor chasing but sexually excited, flounder along the surface only partly submerged for as long as half a minute at a time.

Even stranger than the dual penguin races of the divers are the topsy-turvy swimmings which they perform when sexually excited (Selous, 1927). Either singly or together they throw themselves on their backs with head and neck submerged and legs kicking wildly in the air as if they had been shot; and thus inverted they swim just below the surface. They emerge only to go through these odd antics once or twice again, and then fly rapidly over the loch. On some occasions at least it is the female which starts these aquatic somersaults. Somewhat similar aerial performances have been recorded. Pairs of courting terns glide and side-slip together (Marples and Marples, 1934). It is not uncommon for snipe to fly upside down, the heron has been known to loop the loop, and in display the bluethroat turns complete somersaults (Naumann, 1822–53). Wallace's standard-wing bird of paradise does a remarkable back somersault from his perch, landing with closed wings (Friedmann, 1934a). Other paradise birds such as Prince Rudolph's (Crandall, 1921), the Emperor of Germany's (Wagner, 1938) and the king bird of paradise (Ingram, 1907), as well as some parrots and the green oropendola display in an inverted attitude.

The simultaneous dual execution of the divers' antics deserves emphasis. We see how, by mutual behaviour, the emotional level of the pair of birds is raised and maintained; and a survey of the various ceremonies leaves a vivid impression of the amazing degree to which two birds may be organically and psychologically attuned to one another, so that a very high degree of co-ordination is reached—even in the most extreme eccentricities. It is noteworthy, moreover, that with red-

PLATE XIII

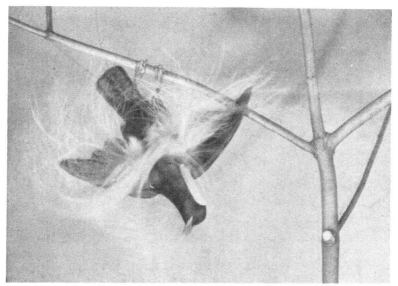

page 130

EMPEROR OF GERMANY'S BIRD OF PARADISE
Inverted display.

page 130

GREEN OROPENDOLA
Inverted display.

PLATE XIV

page 132

KING PENGUINS: ECSTATIC ATTITUDE
AND BOWING DISPLAY

page 61

MALE COMMON TERN DISPLAYING TO BROODING
ROSEATE TERN

The display and habits of the two species are sufficiently
similar for them to inter-breed successfully.

throated divers coition is not the sequel to a series of sexual ceremonies but that it occurs after a period of sexual quiescence.

I have already referred to the mutual greeting ceremonies of whooper swans and the extreme reciprocity of their actions; the ceremonial of mute swans is cruder and does not include vocal duets. They lay their necks across each other and dip them in the water. Whooper swans not only co-ordinate their movements to a remarkable degree but also their calls. Primitive as these duets may be from the musical point of view, the co-ordination of rhythm is a considerable achievement. Without doubt it is to the whooper swan that the Abbé Arnaud referred when he described the duet performed by two wild swans which made their home on the lake at Chantilly. After chasing and mauling a goose,

> Alors les ailes étendues, le cou droit et la tête haute, le cygne vient se placer vis-à-vis de sa femelle, et pousse un cri auquel la femelle répond par un cri plus bas d'un demi-ton. La voix du male va du *la* au *si bémol*; celle de la femelle du *sol dièse* au *la*. La première note est brève et de passage, et fait l'effet de la note que nos musiciens appellent sensible... ils ne chantent jamais tous deux à la fois. (Buffon, 1770–83.)

Whooper swans, however, as my observations show, may trumpet simultaneously as well as reciprocally.

There is a certain resemblance at times between the pattern of the whooper swan's ceremonial and the performance of red-throated divers. The reader may judge for himself by comparing a description of these latter birds with preceding notes on swans:

> It was happily while they were thus observable through the glasses that they broke twice, at intervals, into a clamorous duet of strained clanging cries, extremely loud and powerful, bursting out suddenly and simultaneously, as in a sort of frenzy, and continuing for two or three minutes—at least it seemed so. These outbursts were not, as before, ushered in by the melancholy wailing note, which is now absent *in toto*. In thus clamouring the birds stretched their necks straight out in front of them all along the water, as they

had done on other occasions, but I now noticed that, instead of the head and neck being all in one straight line, the former was raised a little above the water and then bent sharply down again. This gave them a strange wild look with which the clangorous cries seemed in harmony. They first swam towards each other, then followed one another in line, and finally went side by side clamouring all the time in this way—a sight for sair een. That they felt intensely, whilst going through with it, seemed evident....Each time, after this concert, there was a moment or two of what seemed the awakening of sexual activities. (Selous, 1927.)

The sexual display of the little grebe also has affinities with this type of performance. The male and female swim towards each other and when *vis-à-vis* utter a tittering duet with their necks stretched upwards in an attitude strongly resembling that adopted by whooper swans. In this position the patch of lemon yellow skin at the base of the mandibles stands out vividly against the rufous cheeks and dark throat (Huxley, 1919; Hartley, 1937). Western grebes come face to face and wave their graceful necks in a sinuous manner and then perform a concerted rush through the water (Chapman, 1908a).

Two mated king penguins will join in a reciprocal duet. There is a ludicrous air of extreme punctilio about the whole affair. As you watch one of the birds it becomes evident that something is toward. He (or she) draws himself up to his full height with great deliberation, stretches his neck to the utmost, brings his flippers smartly to the front, points his long beak to the heavens and emits a series of sonorous trumpet notes. Then, as the last call fades away, he makes a solemn, abject bow, and thus stands with his beak pointing to his feet in an attitude of devotion. After a moment or two he relaxes and returns to his normal attitude. The bird has the air of dutifully executing some stately, all-important ritual (Gillespie, 1932). Two birds will stand close to one another performing their reciprocal duet. There is no envious competition; Darby and Joan scrupulously observe the correct etiquette. Neither interrupts the other nor breaks in upon the impressive ceremonial. Each in turn sounds his or her horn, salaams and relaxes. Gillespie remarks that there is 'a delicious air of formality and correctness' about the whole performance. I have watched two isolated birds, one posturing regularly, the other standing with a bored appearance beside it.

A less elaborate duet is performed by pairs of gulls. They stand on the shore with lowered head and arched neck, uttering a prolonged wailing cry. This 'duetting' may occur also in the air or on the water, and has been noted in connexion with herring, common, lesser and great black-backed gulls (Darling, 1938).

The crested screamer of the Argentine indulges in the practice of musical soaring flight in pairs or parties. The birds mate for life, and the duet which they perform might be said to be a means of expressing connubial felicity, but perhaps we had better say rather that it helps to co-ordinate the organic rhythms of the birds.

The voice is exercised in a kind of singing performance in which male and female join, and which produces the effect of harmony. The male begins, and the female takes up her part and then with marvellous strength and spirit they pour forth a torrent of strangely-contrasted sounds—some bassoon-like in their depth and volume, some like drum-beats, and others long, clear and ringing. It is the loudest animal sound of the pampas and its jubilant martial character strongly affects the mind in that silent, melancholy wilderness. (Hudson, 1920.)

Why should we doubt that when these birds join in concert, sometimes thousands at a time, making the air tremble in a tempest of sound, that they derive aesthetic pleasure as well as psychological stimulation from their united efforts? And how interesting it is that there seems to be the same tendency at work in regard to vocal duets as in the case of *pas de deux* for dual rites to become communal!

Hudson (1892, 1922) furnishes us with another example of a somewhat similar kind—the performance of the Chiloe wigeon. These birds rise in flocks of a dozen or more and sometimes fly so high that they are lost to view. There they float for an hour at a time, alternately coming close to one another and drawing apart. This aerial dance is accompanied by a bright whistling from the drakes and deep, measured tones uttered by the ducks —a lovely harmony. And as if the beauty of these heavenly tones required an additional emphasis, each time the pair come together they slap each other's wings so smartly that even after the birds have disappeared out of sight in the sky the plaudits with which they accompany their concert can be distinctly heard.

The most remarkable duet which it has been my good fortune to hear broke suddenly into the undertones of sound characteristic of the hours of darkness in the Central American rainforest. As the light declined the jungle thundered with the roaring of the howler monkeys, a concerted, awe-inspiring serenade, and just at the tranquil moment when day capitulates to night the chestnut-headed tinamou sounded his enchanted flute—a few deep, indescribably heart-stirring notes floating from the depths of the jungle. Suddenly, from just outside the hut came a long series of hurried, breathless whistles— oo-oo-$oo$$^{-oo}$-$oo$, oo-oo-$oo$$^{-oo}$-$oo$, etc.—*corcorovado, corcorovado,* with a rapid upward inflexion on the *va.* The calls died away in the distance as if the artiste had hurried off. There was a strange urgency about the performance which made us all look up with interrogation in our expressions. Fortunately Dr Frank Chapman who was with us was able to identify the 'song' and explain its peculiar interest. It is uttered, not by one bird but by a pair of wood quail singing antiphonally. He told us that he had seen the two birds, presumably male and female, stand facing each other at a distance of only a foot or two, singing this duet with perfect reciprocity (1929). So excellent is the timing that one would never suspect that two birds were co-operating. The execution, too, is marvellous, for the syllables are uttered with great rapidity—the one contributing *corcoro*, the other, *vado*, with no perceptible break between. Sometimes, however, the tempo, which increases as the duet proceeds, seems to get rather too rapid for the birds' powers, so that they appear to get out of breath.

Another dual performance which I remember hearing during the hours of darkness in the Panamanian jungle is the clamorous outcry of the wood rails. It is thus described by Chapman (1929):

> One morning a pair of wood rails sang a duet just outside my window. The two parts were quite unlike, one being more elaborate and with at least three times as many notes as the other. But they had the same rhythm and harmonized perfectly. It was a quaint performance. I seemed to be listening to an aged couple singing in shaky, quavering voices a song of their youth.

Various owls sing duets. Amongst such is the barred owl of Texas. Its hoot increases in rapidity and rises in pitch up to a climax, and then with a sob sinks to a final moan. 'Who, who, who, who cooks for you—all?' it chants, according to the negroes. This refrain may be rendered as a solo, but is much more frequently heard as a duet. The extraordinary feature, as with the wood quail, is the precision of the timing. The notes of one bird alternate with those of the other to produce the effect of an echo. Sometimes one bird sings its weird song in a hooting tone while the other utters it as a round of maniacal laughter (Huxley, 1919). Eagle owls sing a duet with the male standing upright and the female bowing at every call (Rettig, 1927). When brown owls engage in duets the male hoots and the female utters *kewick* (Stadler, 1932); the effect is represented in the song at the end of *Love's Labour's Lost*. Long-eared owls also sing antiphonal 'songs', and barn owls carry on screeching duets (Moffat, 1940).

Vocal reciprocation or *Doppelschlag* is recorded of cranes. The birds call, *roor, rirr, roor, rirr,*... The deeper note is uttered by one bird (of either sex) and the higher one by the other in such regular alternation that one bird might be responsible (Sieber, 1932).

The bush shrikes of Africa are famous duettists. The Zanzibar boubou sings 'Oh, dear no, oh', the last, softer note being contributed by the female (Jackson, 1938). In various other subspecies the female interpolates or appends a note so rapidly that only by watching the birds is it apparent that two are calling (Moreau, 1939, 1941). How closely interrelated are the two parts of the duet was shown by Brehm (1911–18) for the Abyssinian gonolek. 'Several experiments have proved to me that the two sexes always act together. I have killed now a male and now a female to make sure. When either falls and is, of course, silenced, the other anxiously repeats the call several times.' Almost any month pairs of the East African barbet may be seen performing a song and dance as if the birds were parts of a mechanical toy, the male singing a bubbling song while swaying from side to side, the female bobbing and uttering clinking notes (Moreau and Moreau, 1937). The cock great Himalayan barbet sings *pi-oo* about thirty times to the minute, while his mate repeats her single note in perfect time with the male's notes some four times more rapidly (Osmaston, 1941). The South American cactus wrens, *Heleodytes bicolor* (Fuertes, 1915) and *H. z. zonatus* (Skutch, 1935), also sing duets. The male of the latter species is always ready to join his mate in a song during the incubation period. Other examples may be found in the literature here cited.

As an aesthetic performance the concert given by two male lyre birds and heard by Chisholm (1934) must be accorded a high place:

He sprang to the limb of a tree, perhaps five feet aloft, and began singing again. This medley had scarcely attained its full strength when a second bird, on a mound perhaps fifty yards off, 'tuned in', and the two birds sang more or less in unison. If there was any spirit of rivalry introduced into this vocalism it was in no sense hostile. Both the voice and the attitude of the bird within view seemed to indicate that he was appreciative rather than resentful of his relative's voice. Indeed each bird listened to the other at intervals, and sometimes each took up a half-completed note, with the result that both birds completed the call in unison. Again, when one bird was half-way through an imitation of the wail of a cockatoo the second mocker took up the call and the two voices ended as one.

He also mentions a lyre bird taking up the call of a whip bird and executing it simultaneously. These mimicry phenomena, however, are of different origin and nature to those we have been considering.

This series of examples of vocal duets in bird life, very incomplete as it is, shows that this kind of performance is commoner than one might have supposed, considering that amongst birds the females are seldom endowed with even moderate musical powers. Females of the British starling (Bullough and Carrick, 1940), robin, dipper and American mockingbirds, sing in varying degrees. Viscount Grey (1927) and others from time to time have noticed aberrant singing hen chaffinches. But in the singing birds proper as a rule only the males sing. The female has not to advertise the possession of territory like her prospective mate, nor is it primarily her business to attract a male to her; if she sang, she would run greater risks of attracting the attention of enemies and being killed; and it is apparently biologically preferable that her energy should be expressed in her reproductive functions rather than vocally. It is, therefore, all the more interesting that these duets occur in as many species as they do.[1]

[1] It should be noted that mutuality or reciprocity in some species is carried to the extreme limit, so that in the common tern (Tinbergen, 1931), shag (Selous, 1927), southern cormorant (Kortlandt, 1938), great crested (Huxley, 1914) and little grebes (Selous, 1902–2; Hartley, 1937), and also in the moorhen (Selous, 1905a), reversed copulation occurs. But only in the grebes, so far as is known, is this reversed pairing successful. In the moorhen reversed copulation is the normal sequence to coition in the usual position.

In considering the origin, meaning and purpose of mutual ceremonies stress must first be laid on the significance of rhythm and mimicry for birds. When we see gannets engaged in a series of sweeping bows to one another we feel that the birds in some sense enjoy the rhythm of their actions; similarly, with noddy terns as they bob to one another, crested grebes as they shake heads together and red-throated divers as they execute their dual swimming rites. Pleasure in imitation no doubt enters into such experiences and has played a large part in the evolution of mutual and reciprocal ceremonies. During the winter court-ship of rooks the male stands with drooping wings and expanded

Reversed copulation may also occur in doves and pigeons (Selous, 1902; Carpenter, 1933) and has been recorded of the woodlark after unsuccessful attempts by the male (Niethammer, 1937). Amongst red deer the hind makes as if to mount the stag before he mounts her (Darling, 1937). Centuries ago Horapollo (II, 95), Aristotle (*H.A.* IX, 8) and Pliny (*Nat. Hist.* X, 51) recorded 'paederastia' amongst partridges. Selous observed unisexual coition amongst ruffs, and Whitman (1919) amongst pigeons. Heinroth, in his work on the behaviour of ducks (1911), deals with 'homosexual fixations'. But Tinbergen (1935) points out that a distinction should be drawn between birds which prefer, and those which content themselves with, an inadequate object. In various species of fly males mount males (Bristowe, 1929). Such also occurs amongst the Phalangideae, and male spiders will court other males. So-called 'homosexual' behaviour has been recorded of snakes, lizards and also of apes. Noble, Wurm and Schmidt (1938) found that unisexual pairs were formed amongst black-crowned night herons in the artificial environment of captivity, and that if a male heron enters an unpaired male's territory and adopts a subservient attitude he may become established as the other's partner. In many species throughout the scale of animal life, if a male fails to react to another male with the appropriate masculine behaviour he is treated as a female. It must be borne in mind that in some species with little sex dimorphism the male cannot distinguish between the sexes by their appearance but only by their behaviour when they meet. This is the case with the American house wren (Noble and Vogt, 1935), the European night heron (Lorenz, 1938), the snow bunting (Tinbergen, 1939a) and penguins (Roberts, 1940a). It may also be mentioned that auto-erotism, which occurs in stags (Darling, 1937), apes and monkeys, has also been noted in parrots, ruffs and penguins (Roberts, 1940a).

tail and bows to the female. She acknowledges his gallantry with a slight bow now and again. This tendency for the object of the display to join in is further illustrated by the behaviour of the magnificent rifle bird. An old male was seen regularly, morning by morning, to take up a position on a branch at daybreak; after preening for a quarter of an hour he began to call. Suddenly, when one of the females which attended him appeared, he faced about and shot out his wings, extended so fully that each quill showed separately and the extreme tips met in front of him. Keeping the wings thus outstretched and throwing his head back to display the beautiful body feathers he swayed from side to side, at the same time making a loud 'woof-woof' with the silky feathers. Occasionally one of the females would step daintily in front of him, stretch her wings and sway exactly as he did, copying his every movement. The play usually ended in a wild chase (Ramsay, 1934). A green sandpiper which gave its ground display to a pair of copulating greenshanks followed suit when they took wing for aerial display (Lack, 1938). Cut a few capers in front of captive common or Stanley cranes, and they will sometimes dance in imitation!

Beebe (1926) noticed that a Galapagos albatross would bow to birds other than its mate (as gannets will do), and that it would even reciprocate his polite obeisance, and great crested grebes will go off and posture with strangers (Huxley, 1914). Rhythmic action is infectious, and thus, as we shall see, elaborate ceremonies in which a large number of birds take part have come into being. Moreover, display·may become divorced from directly sexual ends; that which was originally confined to mating has now become an expression of mateyness. Posturing thus may acquire a wider social reference. This is one aspect of these displays, but they are also, and more particularly, manifestations of nervous tension or, conceivably, self-stimulatory. We have seen, for example, that it is impossible to define where sexual ceremonies end and greeting ceremonies begin, but originally and essentially they may well all have been fraught with sexual significance.

Huxley (1923) has pointed out that mutual displays are often 'self-exhausting' and do not lead to consummation. The great crested grebe has specific pre-coitional ceremonies in which attitudes resembling those of coition are adopted and these are quite different from the mutual display. He notices that these ordinary ceremonies are 'in no way representations of coition activities', and yet he maintains that display is, in origin, an adoption of the coition attitude or something resembling it as a result of brain structure when a certain level of sexual excitement is reached. He concludes: 'In any event, the dual origin of the ceremonies as a whole is clear—a general expression of hyper-excitation combined with more or less of the typical coition attitudes.' But this view is not pertinent to bird display in general. We have already seen that dominant elements in display are derived from emotions aroused and movements made during, for instance, the feeding of the young. Not only coitional attitudes but postures and movements employed in feeding, preening and nest-building may all be worked into the display. The student of bird behaviour soon notices that there are two forms of sex display; activities or attitudes naturally and usually preceding copulation, and activities or attitudes which have not this sequel. Kirkman (1937) tries to separate them into two categories: 'The sex reactions of birds, considered as forms of expression, are of two kinds: firstly, those preliminary to the sex act; secondly, those independent of it and sexually significant only in that they symbolise the mutual attachment of mated pairs.' But the displays of birds are not amenable to such tidy classification. It is often difficult to decide how much sexual significance is attached to a particular posture or behaviour-pattern.

The question, Why do birds sing duets? cannot be separated from the larger problem of rhythm, not only of movement but in physiological functioning. The singing of duets is but a specially interesting example of the varied range of mutual and reciprocal activities. This subject must occupy us later, but it may be said here that it is not always easy for two birds to co-ordinate their sexual curves, and quite evidently doing things together helps towards this end.

Ritual in the bird world plays a part not unlike that which it plays in the human sphere. Only the ignorant and superficial would under-estimate the importance of rhythm and cere-monial in our own society. Efforts to expunge ritual from society have always been unsuccessful. *Naturam expellas furcae, tamen usque recurret.* Our craving for rhythm and design in life, becoming embodied in ordered behaviour, finds expression in ceremonial more or less elaborate. The rules of etiquette, parliamentary procedure, naval and military discipline—all these exemplify the extent to which human life demands to be modelled on a pattern. It is unnecessary to stress how deeply the desire for rhythm and order enters into man's emotional life as shown by the liturgies of the great religions. A common ritual is potent in creating and preserving a valuable emotional tone in human society. It has an analogous function amongst birds.

As every close observer knows, birds as well as men have their individual idiosyncrasies, dependent on various factors, temperamental and physiological. Conjugal accord is a matter of mutual accommodation and adjustment as with human beings. The ideal of

> Two minds with but a single thought,
> Two hearts that beat as one,

is not easy of attainment, but the way to it is made smoother when the two individuals concerned, human or avian, unite frequently and enthusiastically in common endeavours and mutual rites. These, as we have seen in the case of birds, vary from greeting ceremonies merely or entirely recognitional or pacific, to definitely sexual activities replete with emotional intensity. In the bird world rhythmic movement is the most readily available means of achieving and expressing rhythm, but music may also play an important part. It may not only prepare the way for pairing but even set the seal upon it. When a gander has driven off a real or supposed rival and been accepted by the goose, they utter together a clarion cry of triumph, initiated by the gander, and this, not the act of copulation, consummates pair-formation. Their union is expressed in this triumphant shout (Heinroth, 1911).

It is noteworthy that some of the species which sing duets hold their concerts at night. It is then that owls mate, wood quails seek each other, and wood rails are active. May it not be that to sing together in the gloom is a more effective bond of union and token of reciprocal feeling than any scarcely discernible dance or posturing? The mutual sexual curve is thus maintained at a common level and Nature's ends are served.

In daylight, too, dual performances play their part, for the birds which can express their feelings in mutual dance and song have an additional means of attaining to, and indicating their possession of, a common level of emotion. When herons spread their plumes in greeting as they meet on the nest, or when the returning gannet bows to or bills with his mate, these actions indicate, and also tend to preserve, a desirable emotional tone between the two birds—perhaps a specially important matter amongst those which nest in colonies as they could easily find other birds at the requisite phase of the breeding cycle if they felt inclined to desert their partners.

As with many birds' habits, what was originally a biologically useful device has become something performed to some extent for its own sake and for the satisfaction and pleasure derived from it. Amongst such one might class the aerial concerts of the Chiloe wigeon and crested screamer. Aerial evolutions, antiphonal singing and the like are indulged in because birds, like ourselves, feel satisfaction in ordered movement and rhythmic sound. Thus are the bonds of biological utilitarianism broken and birds become artists, acting and singing for the joy inherent in doing so.

In some cases, as for instance the aerial evolutions of the crested screamer, there is little doubt that the pleasure in performing together becomes so great that duets do not completely satisfy the craving for rhythm and imitative activity; thus arise corporate singing and soaring concerts in which many birds take part. It is the finely drawn emotional relationship between the male and female which gives rise to rhythmic activities between the pair, and it is as a later development, and through their contagiousness, that the scope of these is enlarged so that a whole flock may co-operate. The reciprocity of stimuli between the flock and the individuals composing it cannot be doubted.

The artistic performances of birds are due to tension. The opposing strains on bow and string enable the arrow to be shot forward; so internal or social tension may, if it be ordered and regulated, produce not conflict and catastrophe but advance in artistry and progress in social organisation. Social reformers may learn from the fowls of the air. The effort to remove all tension from human society is fraught with great danger, for the misguided reformer may destroy the pattern and upset the rhythm of society. He can only shift the stresses and strains from point to point, and perhaps in his ill-considered zeal blindly concentrate them in destructive force at the weakest spot. Herakleitos, at the dawn of Western philosophy wrote: 'Homer was wrong in saying, "Would that strife might perish from among gods and men!" He did not see that he was praying for the destruction of the universe; for if his prayer were heard all things would pass away.' Man's perennial problem is ever to balance the tension in his own personality and society so wisely that the strain may have constructive power and be productive of things beautiful and of good report. What birds have achieved should not be beyond the wit of man. But reason is not enough; emotion is the well-spring of our actions. Would-be reformers must remember the rugged rocks whence we were hewn and preserve the pattern of life and custom which is part of the essential constitution of human nature. Rhythm and design rule the atoms and the stars, the birds and flowers, and man's supreme task and dignity it is to conform to the great Design, the outline of which he, above all other created things, is privileged in some small measure to discern.

CHAPTER XI

THE EVOLUTION OF SOCIAL CEREMONIES

Aquatic social dances illustrating stages in the evolution of social ceremonial
—Aerial analogues of water-dances—The oystercatcher's display and its
significance—The stimulatory effect of other birds' presence—Corporate
flights—The psychological antecedents and foundations of communal
display.

INSTANCES cited in the previous chapter show that there
is a tendency for dual ceremonies to become social and that
some of them merge into corporate activities so subtly that
no rigid distinction is possible. As we have seen, pairs of crested
screamers soar and sing together in vast choirs; Chiloe wigeon
float aloft in flocks, the drakes piping soprano notes, the ducks
intoning a tenor accompaniment and the sexes drawing near
ever and anon to clap wings together. Such performances have
as their basis not only a sexual but a social bond.

The black guillemot and oystercatcher are amongst the
British species which carry out social as well as dual ceremonies,
and as their performances furnish clues indicating how corporate
ritual came into being amongst birds a short discussion of these
activities will serve as introduction to the study of more
elaborate and rigidly formalised ceremonial.

As I have elsewhere described the 'dances' of these birds in
some detail (1940), I shall merely refer here to certain aspects of
the performances which are germane to the present discussion.

Watch a party of black guillemots on one of the rocky bays
which these plump, black and white birds frequent in the
breeding season; their activity is incessant, whether they stand
on rocks billing and sighing in high-pitched wheezy tones,
flying around the bay, or swimming and diving in the choppy

water. Like mergansers they scuffle, plunge and dip their bills. Very soon, however, it is apparent that there is a great deal of system about their doings; indeed we can classify the 'figures' of the performance as follows:

Nebbing. The birds squat in pairs side by side uttering whining whistles while fidgeting with each other's bills.

Pivoting. One bird swims round the other or couples round one another. Billing and sighing often accompany these movements.

Line formation. Several black guillemots float in a line on the water, not swimming line-ahead or all facing in the same direction.

Plunging. A number of birds dive almost simultaneously.

Pursuing. One bird chases another on or under the water, on land or in the air.

Corporate flights. A flock flies around and alights as a party.

When shovelers pivot on the water as many as ninety-six birds may perform together, but there is no apparent co-ordination between the pairs, though, no doubt, one dancing couple serves to infect others with terpsichorean zeal (Nicholson, 1929). In their waltz, during which the drake circles round the duck, both with heads down and half-submerged, the dancers occupy themselves only with their own partners. But where black guillemots are concerned this is not so. True, the couples often revolve in what might seem the spirit of the song:

> If you were the only girl in the world,
> And I were the only boy!

but watch them for only a short time and you will see that they are far from being entirely wrapped up in themselves. The couples drift or swim about nonchalantly and haphazard, so it seems, until suddenly one realises that they have arranged themselves in a long line spaced at almost equal intervals. Or, as we regard several couples pirouetting close to each other, a bird dives, and almost immediately every guillemot in the party plunges and swims about looking like so many giant pied butterflies in the green depths. In these submarine activities the guillemots are not at all concerned with seeking food. And just now and then all the birds in the bay cease their water-dances and fly around together, a score or more twinkling blunt black and white projectiles in a scattered, intent flock circling the bay and returning to twitter, cosset and bicker on the rocks.

The display of some albatrosses appears to be even more social than that of the black guillemot. Their terrestrial dances will be considered later. In the case of the Galapagos, black-footed and Laysan species these are often communal performances and continue throughout the period of incubation and sometimes longer, but the wandering albatross, in which the sexes differ somewhat in appearance, confines itself to mutual display once pairing-up has been accomplished. But far from their breeding quarters these birds huddle in flocks on the sea, billing, bowing, spreading their wings, bobbing their heads and caressing each other with their beaks. They spin in pairs, changing partners from time to time; even yearling birds take part. The effect of one bird starting to dance is to initiate the activity amongst all the others of the group (Murphy, 1936).

The red-breasted merganser indulges in pivoting, plunging and pursuits, but none of these have the elaborately formalised appearance of the guillemots' doings. In comparison the ducks' activities are wild, passionate and unrefined. I use this latter word advisedly because it emphasises the sharp contrast between the behaviour of the two species. There is restraint and system, an impression of the careful execution of traditional rites about the 'tysties' behaviour, but the mergansers appear rather to be expressing crude instincts, primitive urges and uncontrollable cravings. From November until July these ducks may be seen in small parties gaping at and chivvying their neighbours, as has been described in an earlier chapter. Sometimes one will make a dash at another, splattering along the water, or they submerge and chase each other beneath the surface as guillemots do;

sometimes a bunch will fly wildly around together and alight to display excitedly and frolic madly with one another. But however crude the mergansers' activities may be they are not without suggestions of incipient formalisation too. A pair, feeding together, tend to dive approximately at the same time. Unlike guillemot procedure several seconds may elapse before a bird follows its partner. They do not emerge simultaneously and often bob up some distance apart, but there is a rough and ready approximation. Furthermore, when six or eight mergansers fish together there is a tendency for the periods of submergence to coincide (Armstrong, 1940). Clearly these ducks are not unregardful of what their neighbours are doing and are stimulated to dive by seeing them diving. More remarkable as an illustration of contagious behaviour amongst saw-bills is an incident recorded of the close relative of the merganser, the goosander, by Richmond (1939). Soon after a female began to solicit a drake nearly all the other ducks followed suit. Frequently, after a pair or a party have been conducting their gaping and chasing revels, they separate and start fishing, as I have described; but sometimes, it seems, the corporate diving arises out of the display and is at first sportive and only later develops into the pursuit of prey. I have no doubt that the merganser's habits reveal how the more specialised display of the guillemots evolved. The excitement of the sexual pursuit and the infectiousness of diving and feeding have been fused and refined into a communal romp.

Selous (1901), when watching dabchicks, was impressed by the tendency for the sexual chase to become a corporate revel. On 4 February he saw three dabchicks pursuing each other along the course of a stream.

When quite near they all pitch down and instantly dive. The first to come up stops dead still on the water, looking keenly and expectantly over it, his neck stretched rigidly out, his head darting forward from it at a right angle, as rigid as the neck. The instant one appears, he dives again with a suddenness as of the lid of a box going down with a snap, and this other one has seen him at the same time, and dives still more quickly, if that were possible—so quickly that there is just a swirl on the water, the appearance seems part of the disappearance, 'and nothing is but what is not'.

Selous's interpretation of this scene—so like what may be observed amongst black guillemots—is similar to my conclusions drawn from the study of guillemots and mergansers, that there is a tendency for the preliminary sexual chase to become of secondary importance as birds grow enamoured of the romp into which it develops.

The most casual observer of tame ducks will have noticed that at times they are suddenly and unaccountably seized by a strange excitation and rush quacking, diving and splashing about their pond. This aquatic hurly-burly may be seen at least as late in the year as November. Teal, shoveler, wigeon and mallard, as well, no doubt, as other species, act in this strange way. The performance bears a considerable resemblance to the sudden submergence and pursuit characteristic of the black guillemot, though the ducks apparently do not always chase in pairs as the guillemots do. It appears to be a kind of game, and so far as teal are concerned seems to be more frequent from July onwards and to be unconnected with the sexual display. These evolutions consist in a number of drakes swimming in procession round a selected female, who appears to concern herself little with what is going on. But the point of most interest is that the display, like that of the black guillemot, has become, to some extent, a frolic. Boase (1925) writes: 'As in other cases there is a tendency for the more formal display to become almost a game or tournament, for a number of males take part in a water dance where the interest appears to be more in one another than in attracting by direct appeal one or other of the watching females.'

There are aerial analogues of these water-dances, such as the antics of the black-faced ibis:

> The birds of a flock, while winging their way to the roosting-place, all at once seem possessed with frenzy, simultaneously dashing downwards with amazing violence, doubling about in the most eccentric manner; and when close to the surface rising again to repeat the action, all the while making the air palpitate for miles around with their hard metallic cries. Other ibises, also birds of other genera, participate in similar aerial performances. (Hudson, 1892.)

Sheld-ducks, indeed, act in much the same way. They will, in an instant, break up chevron formation as they forge through the sky, swerve and twist downwards with whistling calls for a few seconds, and then level out, find formation once more and continue their flight. When watching this sudden disarray I have been impressed with the resemblance, *mutatis mutandis*, to the commotion when parties of black guillemots dive together. Possibly this confusion occurs when a decision to alight is abruptly checked, though my impression is that it has a more intimately psychological origin, but it is the raw material out of which evolve elaborate frolics such as those of the Chiloe wigeon of La Plata mentioned in an earlier chapter. In the contrasting notes which these wigeon utter we have another point of resemblance with sheld-ducks. There are interesting similarities, too, between the sheld-duck and the black guillemot, such as community parties, swimming and flying about in couples, visiting the nesting-hole together and circumambulation with dipping of the head in courtship; also the display of a brightly hued mouth in rivalry.

These rather frenzied corporate flights are apparently due, in some instances, to the birds' sensitivity to atmospheric changes. White pelicans become greatly excited at the approach of a thunderstorm. Numbers of these majestic birds mount in spirals from their nesting colony to the upper air; then they dive, 'the roar of the air through their stiff pinions sounding as though they had torn great rents in the sky', and when they draw near the earth they zigzag like swallows chasing insects (Chapman, 1908 *a*). Before a storm the bee-eaters of the Bessarabian steppe become intoxicated with excitement. They gather into a flock and rise almost out of sight in a mazy dance, uttering liquid cries, although normally they are rather silent (Haviland, 1926). Thunderstorms, indeed, have strange effects on birds and men—causing a party of long-eared owls to menace a naturalist (Yeates, 1941) and arousing Swinburne to utter 'a torrent of splendid verse' (Dunn, 1904).

As aquatic and aerial activities tend to become corporate performances, so with terrestrial displays. We need go no farther than our own rocky shores to find a bird whose interesting and puzzling frolics show us all stages from solo to community display. The ceremonial piping parties of the oystercatcher have received a great deal of attention from ornithologists (Selous, 1901; Huxley and Montague, 1925; Dircksen, 1932, 1938; Perry, 1938; Armstrong, 1940). Speaking in general terms the 'dance' consists of two figures, the piping party proper and a procession. Two to a dozen birds may participate, though three is the usual number, especially during the months of February, March and April. As the proceedings vary a good deal in detail in spite of their conformity to a general pattern, it will be best to quote the impressions of two observers. Selous, who first noticed and described the performance, wrote as follows:

When one of the male birds—standing near the female—commences thus to pipe, the other one, if on the same rock, runs excitedly up to him, and pushing him out of the way so as to occupy almost his exact place, pipes himself, as though he would do so instead of him. The other, however, is not to be silenced, but standing close by him the two pipe together, throwing their heads from time to time in each other's direction, and then back again, in a frenzy or ecstasy, as though they were Highland bagpipers of rival clans piping against each other, and swinging their instruments as they grew inspired by their strains. Continuing thus to act, the two male birds approach and press upon the female. She flies to a corner of the rock, the two, still piping vigorously, follow and again press upon her. She flies down upon a lower ledge of it, the two pipe down at her from above. She flies from the rock, they half raise their heads, and cease to pipe, then with single querulous notes, and in their ordinary attitude, walk disconsolately about. After some ten minutes the female flies back again. The demeanour of the two birds is at once visibly affected, and they begin to pipe again, though not so vigorously as before. They continue to do so, more or less, at intervals, the third bird (the female) remaining always passive, and never once piping. All at once one of the two pipers flies violently at the other, who flies off and is closely pursued by him. They alight—it would seem together—on the edge of a great rocky slab, but are instantly at some distance apart, looking at each other and bearing themselves after the manner of rivals.

Perry (1938) describes some of the details of the display more fully than Selous:

> With heads stiffly down-stretched and hackles raised, with half-open, vibrant mandibles straightly pointing the ground like scarlet clothes-pegs (John Donne's 'twin compasses'), and broad tails depressed, two birds run side by side at a third bird, piping hard the while: their motion like nothing so much as two old rams charging, hell for leather. Beginning with plangent *peek* or *pic*, the cheery piping swiftly gathers way into a prolonged diminuendo: *kervee-kervee-kervee-kervee-kervee*, etc., etc., which dies away and crescends into a beautiful rippling. The precision with which the piping pair charge and turn together with lightning swiftness is as astounding as that of small waders in simultaneous flight. They do not attempt to drive away the 'pipee' by physical force, but persist only with this automatous running and turning together when short of, or alongside, the intruder: to and fro, to and fro. I am strongly reminded of the haphazard rushes of the displaying ruff, who is as likely to bow to a clod of turf as to a reeve. But this 'ram-piping' of the oystercatcher serves its purpose, and, ninety-nine times out of one hundred, the 'pipee' soon takes to flight, although I have known a pair to pipe at a third bird for fifteen minutes without a break, before achieving the desired end.

The similarity of the oystercatcher's display to that of the spur-winged lapwing of La Plata as described by Hudson (1892) has been commented on by Selous and others. One of these birds will leave its mate and visit another pair in their territory and lead them in a ceremonial musical march; then return to its own holding to receive a visitor later on.[1]

[1] This habit of calling on one another is characteristic of skuas (Selous, 1901) as we have already noted, some gulls (Darling, 1938) and petrels; it is particularly noticeable in the fulmar. Strange Manx shearwaters may be

In all probability the trios are usually composed of two males and a female, and the performance was originally a competition for the female between two rival males. The emotional bivalency of the ceremonies has already been stressed. An observation of Gordon's (1907) reinforces the opinion of other ornithologists that 'piping parties' are a kind of formalised fighting. He saw a pair of sea-pies endeavouring to evict another couple established on a Scottish islet. The intruding cock drove the female from her nest (Gordon does not say how he identified the sexes), and then all four rushed back and forth with lowered heads whistling loudly. There is no doubt that communal ceremonies may evolve out of territorial jealousy. Galapagos mockingbirds posture on the boundaries of their territories, facing each other with flicking tails and wings. The females will join in and sometimes other birds are attracted so that as many as ten are seen together in these posture dances (Venables, 1940). Other mockingbirds, such as the common species, participate in similar displays (Michener, 1935). Darling (1938) describes 'a communal round dance' amongst herring gulls which he believes has evolved from expressions of territorial jealousy. The 'squabbling parties' of blackcaps and social displays of the Bornean magpie robin (Shelford, 1916) may possibly be of a like nature.

Dircksen (1938) has shown that many oystercatchers do not breed until their third year and also that, as other observers have suspected, the pairs are faithful to each other for years and possibly for life. It would thus seem that a piping party is often due to an unmated male seeking to pair with an already mated female. But this is not the whole story. For example, the late Riley Fortune told me that he saw a female walk off her nest calling.

found in the burrow of a mated bird up to the laying of the first egg, but not afterwards (Lockley, 1939). A redshank has been known to visit a mated pair at the nest (Ryves, 1929). It is well known that birds may feed nestlings of other species (Kearton, 1903).

Immediately four others joined her and, putting their heads down, piped together in a ring.[1] This happened repeatedly on successive days. Thus even an incubating bird can enjoy piping parties. It is comparatively seldom that one sees a male courting a female, but in such cases the pattern of the piping display is retained, the female responding vocally to the male.

The 'procession' is a subsidiary feature of the oystercatcher's display, and just as the piping parties are often corporate performances so in the sea-pies' parades we sometimes find more than two birds participating. Selous goes so far as to say, 'There is really no march at all in the proper sense of the word' when comparing the oystercatcher's display with that of the spur-winged lapwing trios. These, according to Hudson, 'keeping step begin a rapid march, uttering resonant drumming notes in time with their movements, the notes of the pair behind being emitted in a scream like a drum-roll, while the leader utters loud single notes at regular intervals'. But we have the observations of others to supplement and correct those of Selous. Richard Kearton (1915) tells us in a description of his stay on Texel:

> For two or three days in succession my daughter and I watched three birds going through the most curious antics at the same time in the afternoon and at precisely the same spot on each occasion. They marched in solemn procession for a hundred yards or more, the two foremost members of the trio with heads bowed and in silence, whilst the third brought up the rear with head erect and talking loudly and volubly all the time. As soon as a given spot had been reached the procession broke up in something akin to the hilarious din of a tavern joke.

I have also seen such a procession with two birds taking part on Texel.

These antics appear to be a developed form of certain modes

[1] Similar behaviour is recorded of avocets (Makkink, 1936). Up to a dozen birds go to meet each other and arrange themselves in a circle with bills to the ground.

of display which we find in less elaboration in various other species; indeed, the procession is not at all an uncommon figure in bird dances. Consider, for instance, Kirkman's (1937) description:

On a patch of sand on the shore, and so outside the gullery, there walked, with precise little steps, a pair of black-headed gulls. One, which I judged from her restrained behaviour to be the female, walked first, with her neck stiffly erect, but her beak pointing vertically downwards; otherwise her attitude did not differ much from the normal. Her mate presented a very queer object indeed. He also held his beak vertically down, but his tail was fanned and somewhat deflected, his wings hung loose, and his plumage was all puffed out. The pair walked one behind the other; they inclined their heads sometimes to right, sometimes to left, and occasionally the male would turn and incline his head to the prospect behind. This performance, accompanied throughout by the discordant 'Kwurping' or 'Kwarring' by one or both, must have lasted several minutes.[1]

If the black-headed gull shows us a more primitive, less formal procession than the oystercatcher, a more elaborate performance is characteristic of the common crane. Several of these birds will walk in Indian file, one behind the other some distance apart, marching a specific 'parade-step' with long strides and thighs showing well beyond the belly-feathers. As they gradually draw nearer they turn their heads so that the bill is pointing backwards and the red crown exposed towards the neighbouring bird; no doubt, a threat attitude. When one stops they all stop and begin 'false-preening' (Heinroth, 1924–33).

This preliminary survey of aquatic, aerial and terrestrial

[1] The attitude with beak pointing down is adopted in many bird ceremonies. Black-headed gulls stand thus after an altercation while they utter a guttural crooning; the amorous redshank walks round the desired one with feathers puffed out and bill pointing to the ground, and the moorhen's posture of solicitation is to stand with head held pointing inwards beneath the body. When young painted snipe are disturbed they stand with ruffled feathers and bill-tip touching the ground, as the adults do in courtship. (Beven, 1913.)

'dances' has shown that all stages may be traced between the simple reactions of two birds to one another and formal corporate activities. The high pitch of elaboration to which these frolics may attain will be illustrated when we pass to a closer consideration of various types of such dances, but here I wish to emphasise the imitativeness of birds as the clue to the evolution of ceremonial in which a number of birds take part.

It is an interesting fact, attested by Huxley and Montague (1925) and others, that the sea-pie performances in which the largest number of actors are involved last longest and are the most vivacious. Similarly, it has been noted in regard to the ground display of Temminck's stint that the presence of other displaying birds tends to increase the duration and vehemence of the antics (Southern and Lewis, 1938). Moreover, there is a curious collective agitation which passes through a flock of oystercatchers as they stand together on the shore. Some flutter into the air, others run up and down, or bathe excitedly or start piping; but in a few minutes the commotion subsides and they stand stolidly until the next outbreak. The same kind of psychic wave sometimes disturbs avocets and other species (Selous, 1927).

A related phenomenon is the sudden unaccountable flight of birds about their nesting colony already referred to in the case of black guillemots. Possibly the corporate activity stimulates the birds emotionally. An observer in a hide amongst nesting terns or gulls experiences a strange sensation when, for no apparent reason, the clamour ceases, save for a swish of wings as all, or nearly all, the birds in sight fly off together in silence. At one moment sand martins are as busy as bees about their burrows; the next they have drifted away like a swirl of leaves in the autumn wind. After young kittiwakes are fledged all the birds of a colony fly out from the cliff together every now and again, returning to the rocks after a brief flight. Puffins carry out mass flights when they come back to their breeding quarters, and again later, on the failure of first clutches (Perry, 1940). Kirkman (1937) records that black-headed gulls make their first visit of the season to their breeding haunt in thousands, and when the birds first nested on Rathlin they came in a flock. Ruffs abruptly cease sparring and, forgetting their rivalry, fly off together. Even mixed flocks of birds are subject to these ebullitions, minor nerve storms or whatever they may be (Selous, 1901, 1931).

It is easy to believe that such collective flights could evolve into formal aerial evolutions such as those of the Chiloe wigeon, and it is no more difficult to find examples of the raw psychic material out of which they have arisen. Bengt Berg (1931), whilst photographing a brooding crane, noticed that it was subject to recurring fits of nervousness, and it is readily conceivable that the effect of such amongst colonial nesting birds susceptible to 'nervous contagion' would be to cause just such flights as are observed.

Chimpanzees, like cranes, are subject to recurring 'spells of excitement', and Köhler (1927) confessed that after six years' study he was unable to account for them. Baboons will join in chorus if one starts amatory grunting (Zuckerman, 1932), just as a hundred or more gulls are stimulated to begin calling by the cry of a single amorous bird (Darling, 1938). There is corporate sympathetic feeling amongst chimpanzees (Köhler, 1927) and howler monkeys (Carpenter, 1934) to the extent that one will go to the help of the other when it gives the call for aid, but if a captive rat squeals with pain its companions will turn on it and kill it (Alverdes, 1935).

Recurrent excitement due to physiological causes, corporate sympathetic feeling together with a strongly marked tendency to imitative action—these are the materials out of which communal flights, choruses and dances have arisen. In the study of living creatures we should not expect to be able to put a finger on this or that and say that it is the cause or explanation of such and such behaviour. We must think in terms of pattern-complexes, not of chains of reflexes, and be content to follow each strand of the design to discover, so far as we may, how it runs through the pattern and what it contributes to it. No explanation of behaviour in purely psychological, physiological or social terms can be complete or accurate. However obvious this may be, we are constantly liable to forget it. Viewed in this light we may say that ceremonial activities, such as those of the black guillemot or sea-pie, are recreational in so far as the birds derive satisfaction from their antics, but not by any means merely recreational; that they are formal, but not merely due to the tenacity of ingrained habit; that they are social, but not just the product of enjoyed contiguity. And they have arisen through the tendency of one bird in the requisite physiological condition to be stimulated into activity, imitative or aggressive, merely by the sight of another bird performing close at hand. A bird is its true self only within society, constantly giving and receiving stimuli, developing individuality and enriching society by its relationship to its companions.[1]

[1] The reciprocal influence of society and the individual is evident also amongst the social insects. In ant colonies groups show a more uniform rhythmic activity than individuals, very active ants act as "catalysts" to raise the activity level of the whole colony and slow workers are more energetic in company with an active companion (T. C. Barnes, 1941, Rhythm of activity in ant colonies. *J. genl. Psychol.* **25**, 249–55.)

CHAPTER XII

THE FUNCTION OF SOCIAL CEREMONIES

The theory of social stimulation—Corroborative evidence—The importance of imitation—Correlation between communal behaviour and size in terns and penguins—Co-operative nesting—The influence of the social factor on the breeding cycle—The tendency to localise display—The advantages of sociability.

W E have seen that birds are essentially sociable and imitative beings, and that in a multitude of species there are indications of a tendency to associate together and act in a rhythmic or formal way. Before proceeding to discuss some of the more elaborate of these ceremonies, let us pause to consider their biological significance.

In the course of his study of gulls breeding on Priest Island Darling (1938) made a series of observations which throw light on the problem of social display. He discovered that large colonies of gulls not only begin laying earlier than smaller colonies but lay their eggs in a shorter period.[1] Consequently the larger colonies are more successful in raising chicks, for the greater numbers of birds are able to give better protection from predators and the shorter 'spread' of the incubation and fledgling periods reduces the vicissitudes to which eggs and young are subjected. In the last resort, therefore, success in breeding depends, to a considerable extent, on the visual-auditory stimulation of the birds on each other.

The preceding pages have indicated how important is the influence of one of a pair of birds on the other in arousing it to make adequate response. We have also seen that this kind

[1] He also reports that large colonies of grey seals begin breeding earlier than small ones (1940).

of interaction is not confined to a pair of birds. The character of communal displays suggests strongly that there is some valuable stimulatory effect sustained by all the participants. This is borne out by Darling's researches, for he shows that the birds which he studied—the herring gull, common gull, lesser and greater black-backed gulls—all perform dual and communal ceremonies. The common gull carries out a silent corporate up-flight, the herring gull performs an 'aerial visiting dance', both lesser and greater black-backs soar and wheel in a communal aerial dance with much excited calling. It should be noted that the herring gulls studied had compact breeding colonies and commonly displayed near their nests, but the lesser black-backs, whose nests were scattered, had a special communal courting-ground to which they resorted for display purposes— a state of affairs with some similarity to that which obtains amongst sea-pies. It may well be that 'solo' displaying in the colony or flock, such as is common amongst penguins, gannets, cormorants and fulmar petrels, is not without value in maintaining and safeguarding the emotional tension of the community, as Darling's results, with those of Goethe (1937) and Richter (1939), suggest that the mere sight or sound of other neighbouring birds displaying has a sexually stimulating effect.

Darling believes that social displays have this function for those present irrespective of sex-relationship, and are not merely valuable in synchronising the breeding cycle between male and female. He brings evidence to show that the social impact of one species on a closely related species may accelerate reproduction. Small colonies of lesser black-backed gulls nesting amongst herring gulls lay earlier than the majority of these latter when in their own gulleries. This is in line with observations such as those of Lack quoted elsewhere, and confirmed by other observers, that the sight of greenshanks copulating stimulates green sandpipers to display, though it is desirable that Darling's conclusions in regard to the possible influence of one species on another should be verified.

Although much research is still necessary to determine exactly how and to what extent social stimulation affects the breeding cycle of birds, there is a rapidly accumulating mass of evidence that it is a factor of considerable importance. Various ornithologists have commented on the remarkable simultaneity of breeding in small colonies of American tri-coloured redwings, while in larger communities different sections may be in different stages of the breeding cycle (Dawson, 1923; Neff, 1937). Some colonies are most advanced in the centre and retarded peripherally. Lack and Emlen (1939) found that while breeding was synchronous within three colonies a few miles apart, their breeding periods were not contemporaneous—indicating that the birds were influenced by factors other than changes in the environment acting upon the birds' physiological state. This is also true of Indian porphyrios (Hume, 1890). Moreover, simultaneity of nesting is not characteristic of the American red-winged blackbird, which does not breed in colonies. It has, however, been noticed in colonial icterids such as the yellow-headed blackbird and the great-tailed grackle (Linsdale, 1938; McIlhenny, 1937).

Hoogerwerf (1937) discovered that young black-headed ibises on community platforms holding twenty to twenty-five nests were at the same phase on a given platform, but that on other platforms a different stage might be reached. As a result of their study of the black-crowned night heron, Allen and Mangels (1940) came to the conclusion that in this species flock-stimulation is probably essential to reproduction. Roberts (1940a) found no conclusive evidence that large colonies of gentoo or rockhopper penguins laid earlier or had a shorter period of egg-laying than small colonies, but he noticed that in each group the chicks were at the same stage of development, although the different groups varied in age by more than a week. The king penguins of South Georgia varied greatly in the stage of the breeding cycle reached at the colonies inspected, but the young in each colony were at approximately the same stage. Arctic terns (Bullough, 1942), kittiwakes, guillemots and razorbills also synchronise in their groups, but not in the breeding station as a whole (Perry, 1940). According to Johnson (1941) a colony of nesting northern guillemots consists of a confederation of groups which tend to show concentric formation in their laying and hatching dates. When one bird chooses a site others prefer to nest around it, and the social activity seems to enhance the general welfare even although overcrowding results in the destruction of eggs. Although the huddled birds are constantly involved in fighting, they show no tendency to remove to vacant spaces. Evidently, as our study of sea-pie ritual showed, aggressive behaviour may have psychological and sexual stimulatory value. If so, we have here a clue to the significance of lek displays. The coherence of the breeding unit is indicated in the fact that if the outer birds abandon their eggs the remaining guillemots may progressively desert.

Davis (1940a) has criticised Darling's hypothesis on the grounds that the phenomena may be explained in another way —that in large colonies there is a greater chance of a bird meeting another at the same phase in the breeding cycle, and therefore the big communities are more successful. Whatever truth there may be in this, it does not explain such facts as the stimulatory effect of herring gulls on lesser black-backs nesting in their midst. This writer, however, noticed a striking phenomenon amongst colonies of smooth-billed anis which may have a bearing on this matter of social stimulation, although he does not himself draw conclusions from it. Sometimes a group of these birds does not lay eggs until a new member joins it. The inference is that the extra excitement supplies the stimulation necessary to set the whole process in motion. If this be so, we have here additional evidence of social stimulation. Rothschild (1941) found that copulation rarely, if ever, occurred amongst captive black-headed gulls except after the excitement caused by seeing some unusual bird or object.

Apparently a social factor enters into the courtship of the hooded merganser, for those who have studied these birds report that they never observed display between pairs of birds. The evidence indicates that the presence of two males may be essential for display to take place (Bagg and Eliot, 1933). When watching red-breasted mergansers I was impressed by the fact that a party in which were several drakes would go on displaying incessantly for hours at a time, day after day. I believe that with these and a considerable number of other species aggressive display is at least at times sexually stimulating.

Incidentally, attention may be called to the fact that aggressive activities are sometimes based on social instincts. Social defence reactions occur amongst jackdaws (Lorenz, 1938), all the neighbouring red-billed blue magpies try to intimidate an intruder at a nest (Dodsworth, 1910), and when only one or two pairs in a colony of common gulls have eggs or young many birds join in mobbing flights (Yeates, 1934). Some of the avian examples of 'mutual aid' given by Kropotkin (1904) will not bear scrutiny, but there can be no doubt of its occurrence.

In those species in which communal stimulation is believed to occur, the tendency to imitative behaviour is strongly developed. In herring gulls, for example, almost every action from preening to copulation may be initiated imitatively (Goethe, 1937), and kittiwakes will even forage for nesting material in parties (Perry, 1940). It is possible that the performance of imitative actions may cause changes in a bird's emotions as well as its internal state and so have a stimulatory effect on the whole organism. Certainly the importance of the imitative element in bird display can hardly be over-estimated. The infectiousness of yawning amongst human beings suggests that amongst birds imitation may be below the conscious level, and the pathological condition known as 'Arctic hysteria' in which a company of soldiers will commence mimicking their commanding officer shows that the factors which arouse corporate imitative behaviour are not confined to subhuman creatures.

The fact that more birds than is actually the case do not avail themselves of the psychological advantages accruing to those which nest in colonies in no way weakens the theory of corporate stimulation, for it is readily apparent that colonial nesting would be highly dysgenic for many species which might be destroyed *en masse* by their enemies; or, by competition with one another, might make too great demands on the available food supply. The shrinkage of territory which colonial nesting involves can only advantageously occur in those species for which, in the ordinary course of events, there is an adequate supply of food for the community within range.

Bullough (1942) has pointed out that in some families there appears to be a correlation between increase of size, prominence of communal display, and weakening of territorial defence. In the Arctic tern the defence of territory continues even after the young are fully fledged, but the young of Sandwich terns congregate in a 'crèche'. A further stage is illustrated by the Adélie penguin, a bird much given to corporate imitative display; the 'crèches' are guarded by one or two adults, and apparently the birds do not confine themselves to feeding their own young. Sociality of this kind reaches its apogee in the emperor penguin in which incubation of the egg is shared by any bird which can secure it (Cherry-Garrard, 1922). It may also be suggested that lek displays, such as those of the blackcock and ruff, secure for the birds the advantage of corporate stimulation without involving the dangers of colonial nesting; but it will be shown later that physiological and other factors have to be taken into account in attempting to explain either colonial nesting or lek display.

Colonial nesting is carried to its logical conclusion by those birds which combine to construct and share a communal nest such as the monk parakeet of southern South America. The nests may be as much as nine feet in length; each pair has its own compartment, but the whole flock are on amicable terms. During the winter each nesting hole is occupied by five or six birds, presumed to be the original pair and their offspring (Friedmann, 1927; Steinmetz, 1931).

The smooth-billed ani of the American tropics builds a communal nest. One specimen which I investigated had the eggs of several birds in a shallow heap on the floor of the clumsy structure. The birds work, not as individuals but as pairs in building the nest, and there is remarkable harmony in all their dealings with one another. All take turns in incubation and share the feeding of the nestlings (Chapman, 1938; Davis, 1940a). The habits of the groove-billed ani are similar (Skutch, 1935). Another member of the Crotophaginae, the white ani, sometimes nests collectively and occasionally the pairs of birds in the flock breed separately but simultaneously (Davis, 1940b). This is highly interesting, suggesting as it does that the synchronisation of sexual rhythm sustained by several pairs of birds is a preliminary stage in the evolution of corporate nesting. Two Australian species, *Pomatostomus temporalis* and *Struthidea cinerea*, are commonly called 'Twelve Apostles' or 'Happy Families' from the fact that several birds co-operate in building the nest, though in the case of the first-named apparently only one lays in it. Small bands of *Yuhina brunneiceps* construct the nest and tend the young (Yamashina, 1938). In Nigeria auxiliary long-crested helmet shrikes help with nest-building (Clarke, 1940). Year-old non-breeding Central American jays assist the older birds with the nest, and young birds of the species *Psilorhinus mexicanus cyanogenys* actually help the male to feed the incubating female (Skutch, 1935). It is well known that young moorhens (Finn, 1919) and swallows (Williamson, 1937, 1941) will feed their brethren of a younger brood. Young flamingoes feed each other (Chapman, 1908 a). More than one Australian white-winged chough lays in a nest and the young help to build it (Friedmann, 1935). Supernumerary male black-eared bush tits and banded cactus wrens help to feed the young of a mated pair; in the latter species parents, assistants and young roost together in an old nest (Skutch, 1935). A pair of little bush tits will allow a third bird to take a share in incubating the eggs (Addicott, 1938). A young male helps nearly every pair of Australian variegated wrens to feed their young (Waterhouse, 1939). Half a dozen palm chats of Haiti labour together on the large communal nest with separate apartments, and a number of social weaverbirds set to work conjointly to build a domicile, first constructing the common roof and then each couple concentrating on a separate apartment. The structure may measure as much as twenty-five by fifteen feet at the base and be about five feet in height. The birds combine to keep it in repair and use it as a roosting-place during the non-breeding season (Shelley, 1905; Friedmann, 1930). Possibly amongst the advantages of this type of gregarious nesting there is sexual stimulation of the kind detected amongst gulls by Darling and highly probable in the white ani. Since it is well known amongst aviculturists that three pairs of shell parakeets or budgerigars breed better in a cage than one pair, it is not unlikely that monk parakeets nesting communally are stimulated by each other. The significance of psychological and social factors in bird life is vividly illustrated in the contrast between budgerigars and chaffinches or robins in this respect, for only one pair of these out of several in an aviary will breed (Lack, 1940 *d*, 1941).

A considerable amount of work has been done investigating the optimum breeding density of various organisms. Students of animal aggregations such as Allee (1931) have shown that overcrowding reduces the rate of reproduction in creatures as various as Infusoria, Crustacea and insects. One instance will suffice to show how subtle may be the influences which operate between one organism and another. The development of the young of the sea-snail *Crepidula plana* into male or female depends on whether or not it comes into association with another and larger individual of the same species. If the young snail settles near another, be it male or female, it develops into a male, but if no larger individuals are close at hand it becomes a female. The facts are fully authenticated, but the cause is not understood (Allee, 1935).

Amongst birds a variety of phenomena suggests the influence of the social factor on the breeding cycle. In their survey of the fulmar petrel's increase Fisher and Waterston (1941) show that egg-laying is generally earlier in the larger colonies than in the smaller colonies and takes place over a shorter period. They conclude that this petrel's breeding efficiency 'is partly a function of the intensity of its sociality'. It has been thought that gannets need social stimulation to breed successfully, but this has been disproved by the nesting of isolated pairs on the Saltees and at Bempton. Where numbers are too few for a lek to be maintained, blackgame seem to be unable to re-establish themselves. It is possible that once the stock of the heath hen had fallen below a certain number the extinction of the species may have been partly due to the stimulatory social displays, which were so important a feature of this bird's life, being ineffective by reason of the paucity of birds (Gross, 1928). The attempt made to re-establish ruffs in the Broads some years ago would have had much better chances of success if, prior to their release, the birds had been confined in a pen on the ground chosen and become accustomed to displaying there.

What is known of the effects of mutual stimulation strengthens the belief that oystercatchers' ceremonies (to take only one example of communal ritual) are a formalised version of display from which the birds not only derive intrinsic satisfaction but also mutual stimulation, which in turn excites and regulates emotion. We shall see, when we come to consider the tournaments of the ruff, that there are remarkable similarities between these and the piping parties of the oystercatcher, and that the latter's display is like that of the ruff at a lower stage of evolution. There is the same intent raptness, the same atmosphere of formalisation, the same inclination to keep up the performance over a considerable period of the year, the same habit of standing with bowed head and beak nearly touching the ground, the same mixture of enmity and amorousness; and it is highly significant that the oystercatcher tends to longer periods of display as the number of the performers increases just as the blackcock's display is most vigorous at the larger leks (Lack, 1939b). Obviously a ceremony which has both sexual and hostility elements must also, inevitably, if it is to maintain itself, be conventionalised into a social performance. It has been noticed that when a male oystercatcher starts piping before a female, if another or others join in they are at first inclined to jealousy, but the pleasurable mutuality of the party seems to overrule incipient hostility (Huxley and Montague, 1925). Selous (1927) records that when two males piped together to one female 'either they did not attack each other at all, or only after a very long musical struggle, in which they seemed to find mutual enjoyment'. This may be construed as the inhibitory power of social ceremonial supervening over bellicosity, and we can readily realise that, developed sufficiently, it would prevent duels to the death and dangerous mêlées. We shall find that this is precisely its effect amongst ruffs.

Oystercatchers have not definite 'courts' for their antics such as are characteristic of lek birds, but none the less there is a significant tendency to localise their parties.[1] Kearton's account of the performance at the same spot on Texel Island day after day has already been mentioned. Observers of flocks of non-breeding birds find that they frequent particular spots. Ingram and Salmon (1934) say, 'some of these flocks (in May and June) must have numbered several thousand, and they could always be found in the same place'. These writers claim that the birds are seen to be immature non-breeders when closely examined, though this disagrees with Huxley and Montague's opinion that the birds are on holiday for a space from their domestic duties. Huxley and Montague observed that piping parties and even copulations occur in the flocks. The oystercatcher differs from most of its congeners in maintaining a sexual tone when in flocks. The assumption that these gatherings are composed of individuals off duty is not justified, although there is no doubt that mated and incubating birds join in the convivialities. It is known that non-breeding oystercatchers may be either year-

[1] Many creatures besides birds localise their social and other activities. It seems as if the behaviour-pattern needs a familiar territorial setting to be psychologically satisfying. Thus red deer have their traditional playing places (Darling, 1937) and huanacos, the ancestral cemeteries to which they resort when death approaches (Darwin, 1845). Outside the breeding season pairs of dabchicks repair to special areas in the breeding territories to perform trilling duets (Hartley, 1937). In Siberia sexually immature geese resort for their moult to the already overcrowded breeding grounds, although there are plenty of places available, apparently equally suitable and safer (Haviland, 1926). Our own experience shows that routine and familiar surroundings are congenial to mental activity and a sense of well-being. Even the most unpleasant experiences are often more endurable when our environment is familiar.

lings showing a few sperm in the testes or adults whose testes resemble those of breeding adults (van Oordt and Bruyns, 1938). It is probable that a correlation may be found between these facts and the peculiar display behaviour of the oyster-catcher; they should be borne in mind when the *raison d'être* of lek displays is discussed later.

It is clear that where, for one reason or another, there are 'supernumerary' birds we have a situation conducive to the evolution of peculiar, often communal, ceremonies. There is much to indicate that Nature tends to direct pugnacity into the peaceful channels of formal ceremonial, and we shall see, as we study the remarkable behaviour of the ruff, that it is an example *par excellence* of the safeguarding of species from the self-destruction which would result were there not pacific channels into which thwarted emotions are able to escape.

Limitations are set upon sociability, as we shall see, by the laws of social precedence and territory, which operate to further the welfare of the race, but birds, like men and apes, are norm-ally sociable beings. Kaspar Hauser, reared apart from human society, became neurotic, so did chicks of the domestic fowl isolated from their fellows (Vetulani, 1931; Brückner, 1933); Köhler (1927) remarked that 'a chimpanzee kept in solitude is not a real chimpanzee at all'. Social intercourse is not merely an amenity for men, apes and birds; it is a necessity, deprived of which the animal tends to become abnormal. It may be said of birds as Hobbes did of men that without society their life is 'solitary, nasty, brutish and short'.

CHAPTER XIII

THE SOCIAL HIERARCHY IN BIRD LIFE

Social hierarchy amongst birds and animals—Various types of dominance —Peck-dominance and peck-right—Pecking-order between male and female—Mitigation of despotism amongst strongly social birds—Development of despotism—Relationship between position in the social hierarchy and possession of territory—Influence of the endocrine secretions on pecking-order—Dominance amongst reptiles and fish.

BEFORE proceeding to discuss other avian social activities something must be said about the order of social precedence or 'pecking-order' which is a widespread phenomenon amongst birds. Thorleif Schjelderup-Ebbe (1922 a–35 b) has shown that domestic poultry are organised in a hierarchy varying from the dominant bird which arrogates to itself the right to peck every other bird, down to the hen whose unfortunate lot it is to be pecked by all its companions. It has now been discovered that a social order is characteristic of many species. Between any two birds of each species of those examined there exists the relationship of dominance and subservience. This has been proved true of the sparrow, wren, flycatcher, finches, tits, warblers, wagtails, various members of the thrush and crow family, woodpeckers, owls, pheasants, gulls, swans, cranes, storks, herons, curlews, geese and ducks, so that there is no doubt of the wide applicability of the principle of despotism in the bird world. Lizards (Evans, 1936), seals (Katz, 1926), mice (Uhrich, 1938), monkeys and apes (Bierens de Haan, 1929; Maslow, 1934, 1935, 1936) are other animals known to have a social hierarchy.

Exceptions to the rule occur. Groups of smooth-billed anis live so harmoniously together that despotism has not been observed (Davis, 1940 a), and Lorenz (1935) claims that social rank does not exist amongst herons, cormorants, gannets and other colonial-nesting sea-birds. We shall see that this statement needs qualification, but divergent expressions of opinion are not surprising in view of the fact that comparatively little is known of the social hierarchy amongst wild birds, and differences arise owing to varying interpretations of the principle. Ultimately, when our knowledge is greater, it should be possible to classify dominance into types, with distinctions in regard to factors in the environment. It may be true, in a sense, to say that there is a triangular pecking-order between crane, flamingo and pelican, or that the lapwing despotises over the crow (Katz, 1937), but obviously in such instances of the relationship of a species to another species circumstances are so different from those obtaining between members of the same species that it is confusing to classify such various types in one category. Dominance, moreover, is relevant to territorial status. Also, within what may legitimately be called social hierarchy the dominance-relationship between a mated pair of birds is different psychologically, and in regard to the biological ends to which it ministers, from the relationship in a social group, such as a hen-run.

Amongst poultry usually the first meeting decides which of two birds is to 'rule the roost'. On this occasion one of three things may happen: (1) They fight and the defeated bird becomes the subordinate in future. (2) One bird becomes frightened and thereby accepts subservience. (3) Both birds are frightened and the bird which first conquers its fear becomes dominant. The course which affairs take depends on circumstances and on the species of bird involved. Amongst pigeons, for example, it is not so much the first 'scrap' which counts as the result of a long series of settling-in altercations. The newcomer takes longer than a 'new boy' in a poultry-run to accept subjugation (Masure and Allee, 1934 a). This type of social order based on the outcome of a number of conflicts is sometimes called 'peck-dominance' to distinguish it from the type in which rank in relationship to another bird is established in the initial combat. This has been styled the 'peck-right' type. It has been found that injection of a form of male hormone (testosterone propionate) into ring doves is followed by an increased development of a 'peck-right' form of dominance (Bennett, 1939, 1940).

It is not necessarily the strongest bird which becomes despot. If a bird which dominates the flock is confronted by a new bird when it is out of sorts, it may become subordinate to it although the newcomer is a weaker bird. Thus of three birds, *A*, *B* and *C*, *A* may be despot over *B* because of its strength, *B* over *C* because of its courage and *C* over *A* because of circumstances favourable to *C* at its first meeting with *A*. Psychological factors as well as physical strength play their part in determining a bird's status. Naturally the bird lowest in the social scale which submits to pecking by all the others and does not dare to retaliate on any of them is liable to fall into ill health; it can usually be identified by its unkempt plumage. To have a reasonably secure and happy position in society a bird must not be very low in the scale of social precedence. Dominated birds may be so adversely affected as to lose their seasonal sexual (or epigamic) adornments, as in the case of the Asiatic white pelican mentioned by Steinbacher (1938). Once dominance is established, it is maintained in many species by the use of a threat-sound and the punishment of any bird which shows signs of revolting. Insurrection amongst domestic fowl is not often successful, as the rebel seems unable to escape from a feeling of defeatism and on such occasions the despot, furiously angry at its precedence being questioned, fights with unusual energy. If a bird which is despot over two others sees them fighting, it usually feels impelled to participate. It cannot bear to see a subordinate in a fighting attitude and there arises in it a desire to punish the temerarious underling. But amongst other birds such as canaries, for instance, reversals of pecking-order may take place, and it is not true that one bird is invariably dominated by another. The pecking-order is fairly constant in shell parakeets, but in pigeons is more easily altered (Masure and Allee, 1934 b). Subserviency may arise when a bird is at a lower stage than its opponent in the sexual cycle or because the first conflicts took place outside its territory—where any bird's élan is reduced. Time brings readjustment and revenge (Roberts, 1940 a).

Between full-grown birds of various species, whether males or females, social precedence holds, though so long as the sex impulse is active in him the male may refrain from pecking his partner. Investigation into the social order of ruffed grouse reveals that if a bird makes a single blunder in its conduct it is immediately pecked by its fellows and relegated to the inferiority class unless by facing and fighting its tormentors it can prove its mettle. There is no chivalry among them; a male who is not in the final mating stage of his display will kill a female as readily as a rival male (Allen, 1934). Threat behaviour seems to play a part in bringing successful birds into effective breeding condition and may thus have an important function in the display of lek birds.

In many species pecking occurs after coition. This is one of many indications that successful breeding requires the acceptance of subservient status by one of the pair. In the 'labyrinth-fish' type of pair-formation (cf. Chapter XVIII) the display involves the acceptance of 'inferiorism' (Allen, 1934) by the female. Homosexual pairing such as occurs in both sexes of the ruff is often due to a bird accepting the dominance of a member of its own sex and 'reversed pairing' is also sometimes due to female dominance. In some other species such abnormalities are possible by reason of sex recognition being dependent on the member of the opposite sex responding to display with the appropriate reaction. This may not be forthcoming if the bird in question is not at the necessary phase in its reproductive cycle or by reason of illness or any other cause is submissive.

Amongst wild ducks and parakeets the males are dominant during the breeding season, the females during the rest of the year. On the contrary, hen canaries usually dominate the males in the breeding season, but during the other months the situation is reversed (Shoemaker, 1939b). Such changes in the hierarchy may be found to be correlated with alterations in the endocrine secretions. When the female is despot, as obtains in some Passerine species such as the common sparrow, she is liable to be particularly cruel and sometimes delights in power more than in sex gratification, so that no male is able to copulate with her. Exceptions occur, for Tinbergen (1939a) noticed one female of a snow bunting pair who, contrary to the rule in this species, dominated her mate.

There is some evidence that in certain cases dominance in the male is attractive to the female. If a hen is released between two roosters, she tends to move towards the more dominant of the two (Murchison, 1935 a, b). However, Lack (1940e) claims that the aggressive behaviour of robins at pair-formation is not related to dominance in any way.

Noble (1939) urges that sexual dominance should be distinguished from social dominance, but Lack (1940 e) doubts whether we can legitimately use the expression 'sexual dominance' at all. It must be admitted that the dominant-subservient relationship between two sexual partners is not necessarily in respect of sex. Students of our own society are aware that the wife often 'wears the trousers'. It is evident that dominance activities play a varying part in the life of different species according to the stage in the breeding cycle, the mode of display and even individual temperament as well as physiological condition, and that in some cases dominance is applicable only to certain phases or aspects of a complicated relationship. Thus Nice (1938) describes how the male song sparrow, concerned lest his mate should desert during the betrothal period, constantly 'pounces' on her, but she admits (1939) that in many little everyday encounters the female dominates. Possibly the 'charging' of male pigeons, the sudden darts and dives of Montagu harriers and greenshanks on the females during aerial play, the courtship flight of Baird's sandpiper a few feet above the female, and other similar activities such as the 'sexual pursuit' so common amongst birds have dominance implications.

In birds that are naturally strongly sociable despotism becomes mitigated. For instance, the despot in a colony of jackdaws is jealous only of the individuals nearest to him in the social order. Others give way to him and there is no persecution. The result is that birds low in the pecking-order are not done to death. Indeed, there is a device by means of which the bullied and their nests receive protection from the whole community. If a jackdaw is beaten by a rival during the breeding season he flies to his nest and, if further pursued, utters a characteristic cry. The birds of the colony immediately rush to the rescue and the bully is attacked if he does not give way; but usually he joins in the chorus, performing the same ceremonial movements, ducking his head and spreading his tail feathers—a beautiful example of the contagiousness and beneficial formalisation of ceremonial (Lorenz, 1938). Incidentally the jackdaw's submission reaction mentioned earlier, which consists in turning the grey nape to the aggressor, may be contrasted with a similar gesture having an opposite significance—the crane's display of its red crown as a threat.

Very young birds do not manifest despotic proclivities, but these soon develop even though the chicks have no adults to set them an example. Sometimes when older birds dominate the younger the order of precedence does not alter when the younger brood becomes adult. An experiment in hand-rearing black-headed gulls revealed the following state of affairs: A bird which happened to be one of the first to hatch became dictator over the others and retained his position even when these outgrew him. Strangely enough, he 'adopted' the bird which was bullied by all the others so that they became inseparable companions and gave promise of mating. It was found that at the age of six weeks the young gulls already began to associate in pairs, and a bird which lost its partner accepted a ball of cotton wool as a substitute. It was bullied by all its fellows (Rothschild, 1941). Amongst birds there are seldom real friendships between members of the same sex. Of these, friendships between two are most frequent, between three occasional, and between more than three unusual, though certain finches exhibit such relationships and will sit side by side close together on a twig (Friedmann, 1935). Gould's manakins, although the dominant-subordinate relationship is observed amongst them, will sometimes sit as close together as their strict territorialism will permit (Chapman, 1935).

A study of young American black-crowned night herons in captivity showed that the 'pairing-up' of immature birds is related to the appropriation of territory. Very soon the young birds begin to establish and defend their separate preserves. As other birds attain the power of flight they leave the nests and join these somewhat older birds—and thus the first pairs are formed. Some of these are unisexual, but all the pairs are as well defined as those of breeding adults. In every case one bird has to accept dominance by the other, and this relationship is manifested in the ceremony described in Chapter IX; the subordinate bird lowers its head and waves it back and forth with the bill directed towards the side or feet of the dominant bird, at the same time making a clicking sound with its beak. The other bird answers, usually keeping its head higher than its mate. These overtures are frequently repeated. The method of pairing which has already been described is quite different from that which obtains amongst juveniles in captivity. The reader should note, however, that Allen and Mangels (1940) came to the conclusion that a pecking-order in this species is entirely a cage-bird phenomenon. It is of interest to remember that amongst people at the food-gathering stage of culture an unselfish communism exists, but that in the crowded civilisation of cities a pecking-order is clearly discernible.

There is a close connexion between the possession of territory and despotism, but the territory system is admirably adapted to prevent dominance being carried too far. In the artificial conditions of captivity the despot amongst blackbirds may slaughter all the others confined with him (Tinbergen, 1939a). Connected with this is the fact that where a pair of territorial birds such as whooper swans are constrained to remain on a lake dominated by another pair they do not breed. Instances of this kind have come under my own notice. On his holding the male song sparrow tries to bully every bird of his own size. He asserts his dominance by vigorous and incessant song. But a bird without a territory suffers from a feeling of inferiority. When a mature male, arriving from the south, finds his freehold occupied he takes up a servile, supplicatory posture, like a young bird trying to stake his first claim. He fluffs his feathers and holds one or both wings straight up into the air. His song is rapid, subdued and often incomplete. The owner watches him silently, perched with shoulders hunched up in a menacing attitude, dashing ever and anon at him and driving him off; but he is liable to return, and at length fights take place, after which each bird sings loudly. But if the newcomer is an old bird returning to his previous year's freehold he is able to domineer after the first fight or two—not by reason of superior vigour but because of his greater determination and confidence. Eventually both may settle down in territories, as song sparrows at first claim more land than they need and territories are compressible up to a point, but there is a minimum allotment for various species, and in the case of the song sparrow this is two-thirds of an acre (Nice, 1937). Chapman showed that dominant manakins do not dare trespass beyond a certain point in a submissive bird's territory. Lorenz (1938) points out that the chief characteristic of territory is that any individual fights with more vigour if the combat is decided in his own territory, and that the resolution with which the fight is carried on increases as the bird nears its centre. Pugnacity increases as the size of the holding diminishes. Birds that are subordinate in neutral territory become dominant in their nesting territory and captive males are much more likely to win a fight in their home cage than in a strange one. In establishing their territories snow buntings will chase each other back and forth over the boundary, pendulum-fashion, for an hour on end. When one bird reaches a point sufficiently far within his holding his 'courage' reasserts itself and he drives back the other into his own ground, until he in turn is inspired to retaliation by his rights as a landed proprietor. Tinbergen, in his account of the spring behaviour of these birds (1939a), says that 'a male in his own territory is undefeatable'. This is only broadly true of birds in general, as sometimes robins will drive out owners and appropriate their domains to themselves (Lack, 1939a), and Howard (1940) gives a vivid description of the eviction of a waterhen from his pond.

In nature despotism does not entail such sufferings as captive subjected birds undergo through being kept in close proximity to their tormentors. Sexual subservience is not detrimental to the subordinate member of a pair, though the social subjugation which is the fate of some birds when artificially crowded together is so disastrous that bullied birds pine away or are killed. Whether in the wild state or in captivity, weakly or deformed birds are usually viewed askance by their fellows who tend to react intolerantly towards them, but the fact that old and decrepit gulls, too frail even to break the shell of a coveted egg, can survive to such an advanced degree of senility in a colony of the black-headed species, indicates that subordinate birds are able to prolong existence in a wild state beyond what would be possible in confinement; for the black-headed gull is quarrelsome and, as we have seen, precocious in developing an order of social precedence (Kirkman, 1937).

A correlation has been shown to exist between despotism, territorialism and the endocrine secretions. Shoemaker (1939 a, b) discovered the pecking-order in a flock of six female canaries and then injected the three lowest grades with a form of male hormone. Besides starting to sing they assumed the three highest places in the hierarchy. According to Tinbergen, when song thrushes suddenly begin to sing in mid-winter they rise in the social scale (Nice, 1939). Roberts (1940a) thinks it possible that relative degrees of dominance between the individuals of a pair may alter from day to day. Shoemaker (1939b) shows that the territory defended by canaries varies in this way and that it is dependent on the stage in the egg-laying cycle which the birds have reached. Noble and Wurm (1940) proved by experimenting on laughing gulls that an androgen is necessary to induce the breeding calls and postures in both sexes, but copulation does not take place unless one bird has been made sexually subordinate by the operation of oestrogens. Other relevant references are given in Chapter XIX. The effect of the ovarian hormones on territory reactions in lizards has been clearly shown by Evans (1936a, 1937). Lizards with ablated ovaries were found to be more aggressive towards trespassers on their territory than normal lizards, indicating that the ovarian hormones have an inhibitory influence on territorial pugnacity. Testosterone propionate induces male behaviour in adult female domestic fowl or in immature birds of both sexes, including territory defence, nest building, male courtship ceremonies, copulation, and later, brooding (Allee, Collias and Lutherman, 1939).

The invitatory ceremonies of such birds as herons and the various appeasing, nest-relief and greeting ceremonies which are now known to be common amongst birds are devices to overcome or mitigate some of these drives. In turn posturing affects the internal condition of the bird. For example, aggressive display to a stuffed bird weakens with repetition, as has been shown by Lack for robins and Roberts for penguins. This may be a mental phenomenon, but probably the activity of the hormones decreases with increasing demands upon them. Immunity results after repeated injections of anterior pituitary hormone (Collip, 1934).

Unlike the sex drive which is cyclic (cf. Chapter xix) the dominance drive, as is natural with a self-preservatory mechanism, is continuous. Noble and Borne (1938), experimenting with the fish *Xiphophorus* in which there is a strict dominance hierarchy as in fowl, found that if the water of the aquarium is cooled the sex drive disappears before the dominance drive.

Physiological and psychological factors are inextricably interrelated in the bird's organism. When, for example, we say that one bird defeats another in combat, his victory is due to the height of his sexual curve and general health and physique, his relationship to his territory, his experiences with other birds, the intact condition of his 'psychological weapons', and so forth. Research shows ever more clearly that although scientific method forces us to analyse the multifarious factors involved in bird activities we must ever bear in mind the conception of life as an intricate design of behaviour complexes.

CHAPTER XIV

THE DANCES OF BIRDS AND MEN

Sociological parallels between birds and human beings—Examples of bird dances similar to human dances—Classification of dance types—Static, ring, line and place-changing dances amongst birds—Participation of onlookers —Resemblances between the dances of apes and men—Emotional springs of the dance—Imitation of birds in human dances—The function of the dance in stimulating and controlling emotional energy.

T HE reader will have noticed when discussing ceremonies in which a number of birds take part that it is almost impossible to avoid employing terms such as are used in regard to human dancing. This is not due to what some would call the incorrigibly anthropomorphic habits of the modern European mind, but is a natural recognition of the fundamental similarities which actually exist between the dances of birds and men and the identity of the emotional sources from which both take their origin. The resemblances between avian and human dancing are the outcome of emotional drives which underlie the behaviour of all the higher animals; and the natural corollary is that we can use the terpsichorean activities of men to interpret those of birds, and vice versa. Let us not be scared by the bogey of anthropomorphism into the arms of the spectre of Cartesian mechanism. It is not anthropomorphism to believe that man and the higher animals have much in common so far as instinct and emotion are concerned, but an acknowledgement of truth scientifically demonstrated. Katz (1937) writes:

> Whereas a comparison between the societies of social insects and those of men has never led to more than vague analogies, the social life of higher animals has lately proved to have much in common with that of human groups. The far-reaching parallels, existing as they do between social groups of higher animals and human beings, have led to the conclusion that many

sociological phenomena, which so far have been considered as typical for human communities, are yet to be judged as characteristic of all socially living animals, including man.

In spite of the remarkable specious resemblances between the polities of men and social insects the psychological gulf between them and ourselves is very great. In this matter Bergson is a wiser guide than Maeterlinck. Psychological science has proved that emotionally there are strong resemblances between man and some of the higher animals. It would be strange if, having this common psychological foundation, men and animals did not show striking similarities in behaviour. The comparison of avian and human ritual—to use a term inclusive of the dance— demonstrates this. When we endeavour to assess the meaning of dancing amongst birds it would be foolish to forget that we are dealing with a type of behaviour which is as characteristic of men as of birds. In this connexion comparisons are illuminating.

There are many other birds besides those to which reference has already been made whose activities have such order and pattern that the terminology of the dance is naturally and spontaneously used in describing them.[1] To avoid the accusation of consciously or unconsciously laying undue emphasis on the resemblances between the dances of birds and men I shall quote the actual words of some of the observers concerned.

My description of the activities of the black guillemots rather minimised the dance aspect of their water frolics, but it will have been apparent that the co-ordination of movement between a number of birds which they exhibit is something much more elaborate than posturing between two birds, such as we noted

[1] Dancing is here regarded as a type of posturing with an emotional basis, but not connected with the satisfaction of hunger. Thus we are not concerned with rhythmic steps used by feeding birds (Portielje, 1928; Perry, 1940) or the ring-swimming of red-necked phalaropes which serves to render visible mosquito larvae (Tinbergen, 1935). Rothschild's (1941) observations indicate that these are inherited mechanisms which may disappear in default of opportunity for their exercise.

in the case of the gannet; though this, too, may well be subsumed under the term 'dancing' when we consider that some human dances are performed sitting, and consist of movements of the head and arms. Darling (1938) compares the guillemot's frolics to human dances in commenting on their manœuvres: 'The tysties have a communal courtship dance on the sea which is not unlike an eightsome reel or a set of lancers as danced by human beings.' But let us pass to descriptions of the dances of cranes—birds which have long been celebrated in legend and folklore for their dancing.

In his *Report upon Natural History Collections made in Alaska* (1887), Nelson writes:

On 18 May I lay in a hunting blind and was much amused by the performances of two cranes (*Grus c. canadensis*) which alighted near by. The first-comer remained alone but a short time, when a second bird came along, uttering his loud note at short intervals, until he espied the bird on the ground, when he made a slight circuit, and dropped close by. Both birds then joined in a series of loud rolling cries in quick succession. Suddenly the new-comer, which appeared to be a male, wheeled his back towards the female and made a low bow, his head nearly touching the ground, and ending by a quick leap into the air. Another pirouette brought him facing his charmer, whom he greeted with a still deeper bow, his wings meanwhile hanging loosely by his side. She replied by an answering bow and hop, and then each tried to outdo the other in a series of spasmodic hops and starts, mixed with a set of comically grave and ceremonious bows. The pair stood for some moments bowing right and left, when their legs appeared to become envious of the large share taken in the performance by the neck, and then would ensue a series of skilled hops and skips, which are more like the steps of a burlesque minuet than anything else I can think of. Frequently others join, and the dance keeps up until all are exhausted.

Incidentally, it is not uncommon for the dances of aborigines to be maintained until the participants are on the point of collapse.

In her novel *The Yearling* Marjorie Kinnan Rawlings gives the following description of the dance executed by American whooping cranes:

The cranes were dancing a cotillion as surely as it was danced at Volusia. Two stood apart, erect and white, making a strange music that was part cry and part singing. The rhythm was irregular, like the dance. The other birds were in a circle. In the heart of the circle, several moved counter-clockwise. The musicians made their music. The dancers raised their wings and lifted their feet, first one and then the other. They sunk their heads deep in their snowy breasts, lifted them and sunk again. They moved soundlessly, part awkwardness, part grace. The dance was solemn. Wings fluttered, rising and falling like outstretched arms. The outer circle shuffled around and around. The group in the centre attained a slow frenzy.

Suddenly all motion ceased. Then the two musicians joined the circle. Two others took their places. There was a pause. The dance was resumed. The birds were reflected in the clear marsh water. Sixteen white shadows reflected the motions. The evening breeze moved across the saw-grass. It bowed and fluttered. The water rippled. The setting sun lay rosy on the white bodies. Magic birds were dancing in a mystic marsh. The grass swayed with them, and the shallow waters, and the earth fluttered under them. The earth was dancing with the cranes, and the low sun, and the wind and sky.... Without warning they took flight. They made a great circle against the sunset, whooping their strange rusty cry that sounded only in their flight. Then they flew in a long line into the west, and vanished.

It is a pity that we are not told how the sexes were assorted in this dance.

Compare this account with a description of the dance of an African species:

In addition to its general grace the demoiselle crane is best known for its habit of holding realistic dances during the mating season, the males executing truly intricate and beautiful dances to the accompaniment of strange cries, while the females stand around regarding their efforts with every appearance of disdain; but it must, I suppose, be held to redound to the honour of the dancers that from time to time one of the females will stalk pointedly away from the circle followed by the male, still dancing, for whom her nod of approbation was intended. (Treatt, 1930.)

An illustration from a third continent will complete this series of quotations. Maclaren (1926) speaks thus of the stilt birds of North-eastern Australia:

The birds, of a kind known locally as Native Companions, were long-legged creatures tall almost as storks, and white and grey of feather; and the dance took place in the centre of a broad, dry swamp, from the edge of which, in a place of concealment, we watched. There were some hundreds of them, and their dance was in the manner of a quadrille, but in the matter of rhythm and grace excelling any quadrille that ever was. In groups of a score or more they advanced and retreated, lifting high their long legs and standing on their toes, now and then bowing gracefully one to another, now and then one pair encircling with prancing daintiness a group whose heads moved upwards and downwards and sidewise in time to the stepping of the pair. At times they formed one great prancing mass, with their long necks thrust upward; and the wide swaying of their backs was like unto the swaying of the sea. Then, suddenly, as if in response to an imperative command, they would sway apart, some of them to rise in low, encircling flight, and some to stand as in little gossiping groups; and presently they would form in pairs or in sets of pairs, and the prancing and the bowing, and advancing and retreating would begin all over again.

We have thus examples of at least three patterns of dancing— male with female, males with males (and female onlookers), and a corporate dance with partners. This latter apparently differs from the displays of shovelers and tysties already described in that the general pattern predominates over the dual evolutions. These types recur amongst other groups. In detail as well as in general pattern there are resemblances between these bird dances and those of human beings. The apparently disdainful female is not unknown at human dances. The Wadegerenko woman in East Africa looks on at the dance which is intended to win her favour with impassive features, a cigarette in her mouth and her hands behind her. In New Ireland the masked dancers trip back and forth before the damsel, who treats their advances coolly or rudely until at length she accepts one of them (Sachs, 1938). The kind of dance which terminates like that of the crane with male and female disappearing together is typical of another bird, Gould's manakin, whose habits will be described later, and is equally characteristic of various primitive peoples such as the Itogapuk of the Amazon, the Tsaloa of the Rio Yapura (Sachs, 1938), and some Indo-Chinese tribes (Granet, 1926). The men and women dance in couples and then retire two by two into the forest. The human dancer may even stand on his head (Stow and Theal, 1905), as Gould's manakin sometimes does, or turn somersaults like the red-throated diver or the Halmahera race of Wallace's standard-wing bird of paradise.

It would, however, lead us astray from our main purpose if we were to attempt the immense task of making detailed comparison between bird and human dancing. What I am anxious to show is that so far as the emotional and rhythmic foundations of the dance are concerned bird and man share like impulses. Similarity of emotional expression indicates similarity in the underlying emotions. It is when the intellectual element enters that correspondences are hard to find. The comparison of avian and human dancing illustrates admirably the psychological heritage which birds and men have in common.

This may best be shown by grouping dances under two headings according to a classification with regard, first, to Motif, and secondly, Emotional affinities:

Classification according to Motif:
  Static.
  Ring.
  Line.
  Place-changing.
Each of these may be with or without:
  Partners, male or female.
  Other dancers, male or female.
  Spectators, male or female, or both.
  Musical accompaniment, mechanical or vocal.
So far as numbers participating are concerned they may be:
  Solo.
  Dual.
  Dual-communal.
  Communal.
In respect of movements:
  Convulsive.
  Mimetic.
  Reciprocal.
  Complementary.

Classification according to Emotion:
    Excited (or ecstatic).
    Erotic.
    Bellicose.
    Sociable.

It would be instructive to work out detailed parallels between bird and human dances according to this classification, but space forbids. The reader may be interested to pigeon-hole the various ceremonies mentioned in this volume according to the above table. He will find, if he compares the dances described in any anthropological study of the subject with bird dances, that there are few without their avian correlates. I will confine myself to citing some examples of definite types under the two main headings of Motif and Emotion.

As an example of the Static dance we might take the solo posturing of the gannet or the extreme of this type, the boat-tailed grackle, which holds himself rigid whilst awaiting the female (McIlhenny, 1937). Some of the dances performed while seated by South Sea islanders correspond to this type in so far as the movements of the upper part of the body only are effective in the dance. Bicknell's thrush performs a rather more elaborate solo dance and seems to enjoy dancing for its own sake (Wallace, 1939). The closely related grey-cheeked thrush dances in a somewhat similar way, according to Olive Miller (ibid.):

His chosen hour was the approach of evening, when, with body very erect and head thrown up in ecstasy, he lifted his wings high above his back, fluttering them rapidly with a sound like the soft patter of summer rain, while he moved back and forth on his perch with the daintiest of little steps and hops, now up, now down, now across the stage, with gentle noise of feet and wings. No music accompanied it, and none was needed—it was music in itself.

So many human dances are of the Ring type that it is unnecessary to endeavour to specify them. It is the oldest form of communal dance. Turkeys have been seen doing a ring dance, running in follow-my-leader style round a tree (Schmid, 1936), and avocets packed closely on an island wheel *en masse* (Aiken and Warren, 1914). But a more definite dance of this type is exhibited by the rose-coloured starling. The male, in a crouching posture, circles the female with short, quick steps, wing and tail quills in constant vibration, and the feathers of crest and throat erected. All the time he sings vigorously. The female is at first passive, but soon begins the pairing note and joins in the circling movement. So they go around together faster and faster. Suddenly the female crouches and the union is accomplished (Schenk, 1929). This type is, of course, an elaboration of the very common pattern of which we have already had a number of examples in the preceding pages, one bird simply circling around the other. The motif is frequent in human dancing. In the Caucasus we find both types. Amongst the Kundur the man sings as he jumps around his girl, who never moves. But in the dance called the *lezghinka* man and woman circle around each other (Sachs, 1938).

At this point the question arises—to what extent are the spectators of a dance to be regarded as participants? Sometimes, both in bird and human dances, they may join in or take part by beating time, clapping or stamping feet. I have space to mention only three bird dances in which the spectators form a rough circle around the dancers.

The first refers to the display of the grey lag goose and is supplied by Darling (1938):

Six or eight birds form what may be roughly a circle ten yards across, or sometimes there will be two groups that distance apart. A gander steps into the arena, lowers and extends his head as if in menace, begins to walk across the display ground, turns his head this way and that and appears to be hissing (but of that I cannot be sure because I am not near enough to hear); he gathers momentum and finishes the distance in a run and with a flap of his wings. He turns round, walks a step or two here and there, waves his head once more and after a few seconds repeats the performance; another will do the same thing, and the geese not taking part may or may not make small honking sounds the while which are just loud enough to reach me. The spectacle is impressive. This type of display seems comparable with that of the blackcock and other birds of his Family. The display of the geese is not of such elaborate pattern, but to me there has seemed little less of punctilio.

Compare the grey lags' performance with the ritual of the Laysan albatross:

We came upon several gatherings of Goonies and a sight that was both beautiful and humorous—their mating dance. Two goonies stand face to face, wings partially spread. A huge circle of their fellows around them carry on an incessant shout and clacking. As graceful in dance as they are clumsy in walk, the two partners rear their heads toward the sky, their necks arching swan-like, then duck to the ground and, rising, touch beak to beak. Another movement of the head then begins, the pair ducking their beaks first under the left wing, then the right, rearing their heads again to the·sky —two dancers matching step for step. This goes on for several minutes, the clacking of the circle reaches a crescendo as the dancing pair increase their rhythm and motion. In one or two of these circles it was noticeable that one of the dancers would stop as the dance progressed and always—the eternal triangle—there was another fellow waiting on the sidelines to take up the dance and steal the partner. (Miller 1936.)

Albatross ritual has been carefully described by Matthews (1929) of the *Discovery* Expedition. He makes it clear that in the case of the wandering albatross the ceremonial is different before pairing from that which continues afterwards:

In courting before pairing several males gather around one female and bow to her, bringing the head down close to the ground. As they do this they utter a harsh groaning sound, and the female bows and groans back at them. After several bows the males open the wings to about half their extent and side-step around her. They then edge into a position so that they are directly facing her and open the wings to their widest extent so that the tips of the primaries are raised above the level of the head and are curved forwards towards the female. At the same time the males raise the head so that the bill points straight up into the sky, and give vent to a loud braying cry. They then close the wings and start all over again. Several males do this at the same time around the one female, but they do not all act in unison so that unless they are watched carefully one gets the impression of half a dozen male birds dancing around the female and going through a series of haphazard actions, but one finds that they all adhere to the same course of action if attention is directed to each in turn.[1]

Such dances are of the pattern typical of many primitive dances in which one or two men or women are circumambulated or ringed around by others. In a cave at Cogul in the province of Lerida in eastern Spain there is a coloured rock painting of the Capsian period (Miolithic) which represents some such dance motif as that of the wild geese and albatrosses. A naked man stands in an arrogant attitude with his chest thrown out in the midst of a group of nine skirted women. This dance seems to have had a long history, for it appears as the dance of the nine shepherdesses around Krishna and the nine muses around Apollo. To-day the Bushmen women dance around a man and the Wanyamwezi bridegroom stands inside a circle of women (Sachs, 1938). A Bushman painting depicts

[1] After pairing and before incubation starts the ceremonies are even more elaborate. Groaning and bowing, the male presents nest material to his mate as she sits on the half-built nest. She returns the salutation and then arranges the material with her bill and feet. He squats alongside and gives a bubbling cry, stretching his head up and braying with bill open at the end of each call. She answers, and after they have nibbled each other's feathers he goes off for further supplies. But after every four or five loads and the usual bubbling, braying and nibbling of each other's heads and necks the female steps down from the nest and facing each other they stretch up their heads and bray with wide-open bill. Immediately after this they plunge their bills vertically downwards amongst the breast plumage and utter a low, inspiratory note. They then lean forward and touch the tips of each other's bills. After this they both keep their necks forward and bend their heads slightly upwards vibrating the mandibles to produce a rattling sound, rising from a low note to a high one. Having repeated these antics several times the male walks sideways round the female working his head while she moves to keep facing him. Now with wings stretched out and head upstretched he rattles his mandibles again. The female answers him, at the same time stretching her wings also. Standing thus they reach out and touch bills, then vibrate mandibles and touch their breasts with bill-tips a score or more times in succession. Finally, they copulate and the female sits on the nest while her mate returns to the work of collecting nest material.

a group of men dancing while women at either side clap hands, probably beating time (Tongue, 1909).

Spectators, human or avian, may assist not only by beating time for the dance but by providing music, though sometimes the dancers provide their own accompaniment. The whooping cranes have already supplied us with an illustration of this. So far as birds are concerned probably the most remarkable performers are the manakins of Central and South America. Bigg-Wither (1878) provides a description of one of these little ballets. Hearing a bird singing in the woods the explorer's interest was aroused, as he had hitherto noticed few song birds in the district he was investigating. His men led him cautiously through the forest to where there was a stony patch of ground at the end of a glade. A number of birds of the size of tits, with blue plumage and red crests, were perched on the stones and surrounding shrubs. One standing on a twig was singing merrily while the others kept time with wings and feet in a kind of dance, twittering an accompaniment. The natives with Bigg-Wither called the tiny artistes 'dancing birds', and the description undoubtedly indicates some species of manakin. Von Ihering (1885) describes the long-tailed manakin of southeastern Brazil as the 'Dansador' or 'dancing bird' and writes thus of it:

The males, at most three or four in number, in the spring perform their dances before the females who sit by quietly and look on. They require for this a thin horizontal or slightly inclined twig upon which they bob up and down. Unfortunately, I myself have observed this beautiful performance only once, and not long enough, as we were observed. According to the assurance of many acquaintances, one of the creatures is said to make the music for the dance, and later, with a beating together of the wings, by a loud sharp *Pfiff* gives the signal for the breaking up of the play. Also, they are said often to use the place regularly for a long time.

Our third type, the Line pattern of dance, is exemplified in the description of the sheld-ducks studied by Turner (1928):

They are at the height of their courtship display in April, when numbers collect in certain quiet places and indulge in tournaments of a kind. Sometimes a double row of drakes will advance towards each other with lowered heads and open bills, arched necks and outstretched wings. The preliminary steps are slow and stately, but gradually the pace quickens. When the two lines meet, the wings are raised and arched, yet pressed close to the body. Then the neck is suddenly thrust forward like a couched lance, and the drake which gets the first thrust home throws up his head vertically. This seems to show that he has scored a point. He then moves his head rapidly up and down with sinuous thrusts of the neck, so that the latter seems to form a series of loops. Sometimes these sham fights take the form of duels, while the rest of the company run about making remarks. But whether the game takes the form of a duel or becomes a general *mêlée*, it seems entirely lacking in envy, hatred and malice, or any uncharitableness. It is a beautiful game to watch and also amusing. The sheld-duck's upper mandible curves upwards, and when the brilliant red bill is wide open as the birds rush squealing at each other their whole expression is irresistibly funny.

This procedure is not unlike a dance executed by women of Palau (Caroline Islands). Their bodies and legs are painted red and they approach each other in two bands gesticulating wildly and brandishing spears. Two or three paces apart they form into rows and sing a monotonous song, swaying their hips the while in a strange motion. The dance represents a battle and concludes with a loud outcry (Semper, 1873).

The Place-changing dance is the type in which the dancers alter their positions in relation to their partners or other dancers. Gould's manakin performs the figure known as 'setting to partners'. The dance is described in Chapter XVI. Partridges also have a dance in which this 'figure' appears. When a number of birds are involved we have evolutions akin to those of the 'Native Companions' of Australia or the fourteen razorbills observed near Priest Island by Darling (1938):

The birds were almost in single file at first and, because their legs cannot be seen moving beneath the water, it seemed as if the movement of the sea itself brought the birds into a group, beaks inward. Their raised beaks almost touched in the centre; the circle enlarged and broke; each pair of birds bobbed and both members came together and held beaks. In this fashion the pairs waltzed round for a few seconds. Then all the birds formed into single file again, headed in one direction, and struck a posture of ecstasy which was maintained for three or four seconds. The beaks were lifted high and slightly open and the tails were raised even higher than normally. The ring and the pairs were formed again, and the state of ecstasy was reached while the members of the pair were facing each other; then more waltzing and the curious single file and the posture of rapture throughout the flock once more. The whole display lasted for about a quarter of an hour, after which the birds began the common occupation of wing-flapping and preening.

This account, with a few modifications, might be a confused description of a folk-dance such as 'Newcastle'. It has the same movements back and forth from the centre, bowing to partners, linking of arms corresponding to the holding of beaks, and the formation of lines by the dancers.

It is interesting to note that this kind of dance in which the dancers hold one another does not appear until comparatively late in human pre-history—the Neolithic period. Before that epoch the dance seems to have been executed by a group of individuals not designedly co-operating in a dance pattern. If the razorbill's dance illustrates a remarkable degree of social co-ordination, it indicates the limits beyond which birds are unable to pass. In place-changing dances which consist of more than a few simple movements in reciprocal relationship there enters an intellectual factor—a certain amount of memorising and thinking ahead. This threshold of achievement the bird is unable to reach. Emotionally and in sense of rhythm man and bird have not a little in common; it is man's possession of reason which sets him in a category apart.

It would be strange if the affinities between humanity and birds in emotional expression could not also be found between man and mammal. There is abundant evidence that this is the case as accounts of dancing amongst apes clearly show. Köhler (1921) gives the following description of his chimpanzees:

In mock fighting two of them drag each other about the ground until they come to a post. The frolicking and romping quiets down as they begin to circle about, using the post as a pivot. One after another the rest of the animals appear, join the circle, and finally the whole group, one behind the other, is marching in orderly fashion around the post. Now their movements change quickly. They are no longer walking but trotting. Stamping with one foot and putting the other down lightly, they beat out what approaches a distinct rhythm with each of them tending to keep step with the rest. Sometimes they bring their heads into play and bob them up and down and with jaws loose, in time with the stamping of their feet. All the animals appear to take a keen delight in this primitive round dance. Once I saw an animal, snapping comically at the one behind, walk backwards in the circle. Not infrequently one of them would whirl as he marched with the rest. When two posts or boxes stand close to each other they like to use these as a centre and in this case the ring dance around both takes the form of an ellipse. In these dances the chimpanzee likes to bedeck his body with all sorts of things, especially strings, vines, and rags that dangle and swing in the air as he moves about.

This is very similar to the round dance of many primitive peoples. Moreover, it is interesting to note that the chimpanzee has a taste for adorning his body for the dance. Birds, too, decorate themselves, for cranes anoint their backs with peat water and mud (Berg, 1931), the great hornbill of India paints itself on beak, casque, neck feathers and crown with an oily secretion from the uropygial gland (Hingston, 1933), and the mot-mot plucks the barbs from its tail feathers, leaving an oval feathered extremity.

In the paper already referred to, Köhler relates that when he appeared unexpectedly a female chimpanzee hopped first on one leg, then on the other. Primitive peoples sometimes hop likewise on seeing a white man. In both cases the reaction is due to tension and fear. Köhler also observed apes not only performing a whirling dance, but combining a forward and rotary movement, such as appears in human dancing.

The round dance of the apes is a social dance, the hop a dance of excitement, but they also have erotic and war dances.

According to R. M. and A. W. Yerkes (1935) the male chimpanzee 'frequently performs a crude dance preparatory to
sexual contact. It consists of rhythmic pounding of the ground
with feet, hands or both, clapping the hands together, and the
striking of surrounding objects, especially those which resound
with blows.' Rothmann and Teuber (1915) describe a male
chimpanzee dancing a sexual dance in triple rhythm close to
several females and also a female performing dance movements
in the presence of a male. The resemblance between these dances
and those of primitive races was very marked. Drumming, it
may be mentioned, has frequently been noticed as an activity
of wild chimpanzees. Garner (1896) claimed that they made
drums of clay and held carnivals around them. His account
may be exaggerated, but the drumming parties are well
attested (Nissen, 1931). Chimpanzees will clap hands when
excited just as children do.

Jimmy, the chimpanzee which used to be in the London Zoo,
began his remarkable dances by banging loudly on a metal door
and ended them with a dramatic leap towards the spectators.
The actual dance steps were sometimes extremely intricate. Not
only was there a cross-rhythm as between fore and hind limbs
but the hands beat once to the feet's twice. After a considerable
period of this dance the ape became quite wrought up and
swung round and round, swirling the straw on the floor of the
cage (*The Times*, 1938a). The dance is shown in the film
Monkey into Man edited by Huxley.[1]

[1] The chimpanzees, Jackie and Peter were taken to see this film and their
reactions recorded: 'Jimmy's dance provided the most striking reactions.
Peter began to shake his head rhythmically from side to side, and then
advanced (pulling his keeper after him by his chain) towards the screen.
There he held on to a bar and looked up with the fascinated air of a small
child gazing at a conjurer. At one moment he turned round and began to
dance himself. Subsequently when the late lamented gorilla, Mok, was
depicted jumping at the audience, Peter retired hastily. From gorillas, the
film passed to Australian natives dancing: this again stimulated Peter to

The gorilla beats his breast in anger as primitive men such as the Aponto of West Africa do in their war dance. A party of these apes has been seen assembling to execute 'what seemed a sort of crazy war dance' (Burbridge, 1928). In the Congo, before the natives go out hunting they imitate the habits of the gorilla, and its movements on being attacked (Young and Smith, 1910). Falkenstein (1879) speaks of a gorilla which clapped its hands and whirled about in a crude dance as if 'drunk with pleasure.' Orang-utans also sometimes invent dances (Yerkes and Yerkes, 1929).

Physiological and emotional needs, sociability, the urge towards the satisfaction of the sense of rhythm, and the inherent tendency for life to express itself in patterns of behaviour, have given rise to similar activities amongst birds, beasts and man; thus it is no mere capitulation to anthropomorphism when those describing bird dances use terms such as 'quadrille', 'minuet', 'waltz', 'pirouette', and 'setting to partners', but a recognition of the resemblances between avian and human dances. Dancing of one kind or another is characteristic of so many creatures from spiders to apes that we are justified in believing that amongst men it is a natural and spontaneous expression of the feelings. The mimicry of animals is a relatively late development characteristic of the stage when man, impressed by the superior strength and *savoir vivre* of the animals, copied them in order to acquire their virtues. To this matter we shall shortly recur.

rhythmic swayings. But when the scene changed to shots of city life he lost interest. The end of the film was then run through again, but without sound. Jimmy's dance now provoked much less interest, the animals merely gazing intently from their seats. This was perhaps to be expected, as the sequence is much less effective to humans when given without sound. The Australian black-fellows dance elicited rhythmic movements of the head from Peter, probably as its rhythm was visually much clearer, as the men were silhouetted against the sky '. (*The Times*, 1938*b*.)

At this juncture we may refer to our psychological or emotional classification of dance types under the following heads:

>Excited (or Ecstatic).
>Erotic.
>Bellicose.
>Social.

It will be seen that each of these types is found amongst the anthropoid apes.[1] It is hardly necessary to stress the fact that they are also characteristic of birds. As an example of an excitement dance might be quoted Hosking and Newberry's (1940) account of the strange behaviour of a partridge on the hatching of the eggs.

The cock, peeping into the nest, saw his first-born, and went wild with excitement. He rushed frantically about the place, banged himself against the old oak, and came back to the nest for another peep as though he could not believe his eyes. Seven times he came back to convince himself the eggs were really hatching, and meanwhile the hen stood to one side and looked on proudly.

Excitement dances grade into the Ecstatic. Earlier pages give instances of these amongst birds, and examples amongst men are so familiar as to make detailed references superfluous. In the rites of Bacchus and Cybele dancing was employed to produce an ecstatic state little short of madness. In Bali and Indo-China

---

[1] It is impossible in this short consideration of the subject to deal with dancing as it is known amongst mammals as a whole, but the following note on the dances of stoats is of special interest as it shows that the animals possess an innate knowledge of their 'art'. The owner of a captive stoat writes: 'At times she will perform her elaborate "death-dance" around an odd leg or ear (of a rabbit)—or merely for her own amusement around some imaginary rabbit if none has been produced for her. In nature the stoat seems to "mesmerise" its prey by performing this dance, and it is interesting to find that young stoats taken from the nest before their eyes are open, and reared by hand, execute the whole elaborate dance by instinct.' (Rothschild, 1941.) Hudson (1913) gives a vivid description of this dance. Cf. also Priestley (1937).

to-day the dancer goes into a trance. Just as the prairie sharp-tailed grouse may fall a victim to the coyote during his amorous ecstasy, so during the Boer war the Bushmen were surrounded and shot down whilst rapt in the performance of their dances (Sachs, 1938).

Erotic dances amongst men and birds range from simple solicitous posturing to such elaborate performances as have already been described. The sheldrakes' dance, and the tournaments of ruff, blackcock and other Tetraonidae which are mentioned later, represent the Bellicose dance type, and the frolics of the tystie or the aerial manœuvres of the whistling wigeon may be grouped amongst Social dances.

Once again, it must be emphasised that these classifications are not rigid, for a dance of one type may contain other emotional elements. The reader by now will need no reminding that diverse feelings such as love and anger may dwell very close together in the heart of a bird. This is also true of man. In discussing the strange combinations of war and love motifs in dances Sachs (1938) says:

> But behind this relationship is the emotional make-up of man. For him the battle impulse and the sex impulse are as closely intertwined as they are for the stag that battles when in heat. The ecstasy of blood and love flow together in life as well as in the dance, which here, too, uses life itself as a model: an erotic dance is often added to the war dance, or in taking part in a war dance women become sexually aroused.

Without this realisation of the intertwining of emotions we can never understand the behaviour of birds, mammals or men.

It is not surprising, in view of such facts as those with which the foregoing pages are concerned, that a close student of humanity, Havelock Ellis (1922), should observe that animal dancing is very human-like in appearance. It is so because the foundations of behaviour are so similar in animals and men. They have like emotional drives, appetites and instincts, and to a considerable extent find satisfaction in the same rhythms and orderly patterns of activity.

The reflex of this is seen in the degree to which man imitates the animals. It would be absurd to hold that the general similarity of avian and human dances is due to anything other than a parallel expression of similar emotions and a common enjoyment of a sense of rhythmic pattern, but it is well known that man often chooses to mimic the animals for his own purposes. Primitive folk regard animals as superior beings, possessed of magical power and able to control the powers of Nature which give or withhold fertility. Thus they imitate them in their dances with a view to achieving one or more of the following desirable aims:

> To secure success over the animals in hunting.
> To foster the increase of useful animals.
> To propitiate the spirits of the slain.
> To appropriate to themselves the powers of the animals.

I would emphasise that man's essential affinity with birds and animals in respect of emotional endowment, which is illustrated on the less reflective plane by the resemblances between man's most primitive dances and those of animals, receives confirmation on a higher level when man deliberately copies other creatures.

In some posturings it is clearer than in others that the imitation is deliberate. One of the oldest recorded animal dances is the crane dance which was performed by Theseus on his return from Crete with the youths and maidens he had liberated. It was a round dance in crane step. There was a 'dance of the white cranes' in China in 500 B.C. associated with a ritual pattern so similar that it is difficult to avoid the conviction that the two were related (Granet, 1926; Sachs, 1938). The Jivaro Indians mimic the cock of the rock in a sensual dance, the women singing:

Being the wife of the cock-of-the-rock,
Being the wife of the little *súmga*,
I jokingly sing to you thus:
Cock-of-the-rock, my little husband,
Wearing your many-coloured dress of feathers,
Graceful in your movements!
I know I am useless myself,
But still I rejoice,
For I am the wife of the *súmga*—
So I jokingly sing.   (Karsten, 1935.)

Another example of a dance adopted from the nuptial activities
of birds is found amongst the Chukchee of Siberia who mimic
the notes and movements of the ruff (Bogoras, 1904–9). The
Monumbo of New Guinea have a cassowary dance (von Vormann, 1911), and the Australian aborigines imitate the emu, the
natives of New Ireland mimic the hornbill, representing even
the feeding of the young, and the Maidu of California have a
rather conventionalised representation of the tree-creeper (Sachs,
1938). Both end in a flapping of the arms indicating the departure of the bird. One of the two principal dances of the
Tarahumare Indians of Mexico is an imitation of the spring
antics of the turkey.

> For the strange behaviour of many animals in the early spring the
> Tarahumares can find no other explanation but that these creatures, too, are
> interested in rain. And as the gods grant the prayers of the deer expressed
> in its antics and dances, and of the turkey in its curious playing, by sending
> the rain, they easily infer that to please the gods they, too, must dance as the
> deer and play as the turkey. (Lumholtz, 1903.)

Both these dances are performed by numbers of men and
women, the sexes keeping apart from each other.

The Kobeua and Kaua Indians of north-western Brazil mimic
birds, beasts and insects in the masked dances at festivals in
honour of the dead, and the creatures impersonated range from
the beautiful blue *Morpho* butterfly to the house-spider and the

dung-beetle, from the sloth to the swallow. The black vulture and the owl are also represented. It is supposed that by copying these creatures their power is acquired by the dancer, who in so doing gains magical influence over the spirits of life and fruitfulness, and thereby brings the blessing of fertility to the village, its plantations and the whole of nature (Koch-Grünberg, 1909–10).

Animal mimicry is not unknown in European dancing. One of the figures of the Bavarian *Schuhplattler*, the *Nachsteigen*, depicts 'mountain cock mating'. The man jumps along behind his partner, hissing, clicking his tongue and clapping, even turning somersaults. Finally, he strikes the ground with one or both hands and bounds or rushes towards her with arms outspread or hanging close to the ground (Sachs, 1938). There is an interesting similarity not only between this dance and that of various species of Tetraonidae but also between it and the sexual dance of the chimpanzee. In many instances the performers of animal dances decorate themselves with the adornments of the animals and birds represented—horns, beaks, plumes and so forth.[1] Probably this is the origin of the blackcock's feathers in the Bavarian highlander's hat.

Bird dances may be performed in all the elements. Man, earth-bound as he is, has to content himself for the most part with terrestrial dances though he does his best to surmount his limitations. Some primitive peoples perform to some extent in the water (Sachs, 1938) and even the air has been conquered as

---

[1] 'A Naga warrior in ceremonial dress is a really imposing sight, and the women's hearts beat faster when the men, flourishing spears and daos (hunting knives) as if setting out on the warpath, rush through the village with wild cries. With their tossing feathers (of the giant hornbill) and the gaily dyed hair waving from clothes and weapons they look like fantastic birds, and like birds in the mating season the males, rejoicing in the glory of their multicoloured spring accoutrement, reduce the females to comparative insignificance.' (von Fürer-Haimendorf, 1938.)

a dance medium by the Mexican Indians in their 'flying-pole' performance.

I believe that the facts which have been adduced show that the dance, apart from intellectualised types, whether magical or social, has roughly similar functions amongst birds and men, and in so far as we are able to understand the psychological foundations of our own dancing we are enabled to comprehend the dances of birds and the mentality of the dancers. Human dancing has been called 'a method of auto-intoxication of the very greatest potency' (Sergi, 1901). There is some truth in the statement that 'a girl who has waltzed for a quarter of an hour is in the same condition as if she had drunk champagne' (Lagrange, 1889), though obviously the quantity of champagne and quality of the partner should be taken into account. Few will doubt that such water-dances as those of grebe and diver are also in some sense auto-intoxicating. For human beings dancing temporarily enriches the texture of life. We may assume that it does no less for birds; and the assumption goes far to explain the *raison d'être* of many a ceremony and remarkable flight evolution.

Sachs declares, 'All dancing is originally the motor reflex of intense excitement and of increased activity', and Crawley (1918) calls the dance 'a translative engine of human energy' and says, 'on the physiological side dancing develops energy and releases it; it promotes tumescence and effects detumescence'. This two-fold reference or bi-valency of dancing requires emphasis. On the one hand, it is a means of working up and augmenting emotion; on the other, it is a means of expressing it and a safety valve for the over-plus. Dancing would not be so universally popular did it not serve manifold purposes. We have seen that a bird may dance or sing when defeated in battle or frightened. So do men dance at funerals as well as at weddings. Belligerency and eroticism may mingle in the dances of birds as in those of men. The dance has the function of a power station both in concentrating and distributing energy, which, if directed in wrong channels or allowed to accumulate beyond a certain point, might cause disaster.

In the light of this conception some of the problems discussed in earlier chapters may be more easily understood and more readily solved, for such movements as 'wing. fluttering' and 'gaping' may be regarded as posturing akin to dancing and interpreted according to the principle of bi-valency. So also may the problem of the similarity between love- and war-displays discussed by Hingston find its solution.

The performance of the war-dance helps to raise tribal morale before the battle. Psychologically, and perhaps even physiologically (Cannon, 1929), the antics have real value, for each participant tends to rouse his comrades' spirits as well as his own as he prances and leaps, brandishing his weapons. The activity intoxicates him and focuses his energies on the feats of valour he believes he is about to perform. He goes forth a brave man, and the posse of which he is a member is the stronger by reason of the community feeling engendered by the common activity. An incidental result is that the hearts of the enemy are more prone to quail as they hear the distant thump of the dancers' drum or see the gesticulating warriors advancing. It is not surprising that birds have developed a great variety of 'psychological weapons' when we realise that a bird's status in society and its very existence often depend on the impression it makes in its first few encounters with others of its species. Effective movements and attitudes are useful in frightening an enemy or stimulating a potential mate—hence the biological importance of the dance.

It is almost impossible to exaggerate the potency of one bird's display on another. A particular image on a bird's retina is sometimes like a magic key starting up the mechanism of behaviour. Click goes the picture; bang goes the reaction. The course of events, however, is often not as simple as this. The bird being courted may not be quite ready for the nuptial rite. A fine adjustment to achieve synchronisation is necessary. Must the suitor wait, idle and helpless? No! he can posture and prance, sing or demonstrate in the air until her desire is aroused and she plays her part in the dance—that great sequence of courtship, procreation and migration which constitutes the dance of life. The female, too, may invite and stimulate the response of the male by dancing before him or assuming a solicitous attitude. What, indeed, is life but a ballet of indescribable complexity and infinite significance? All play their part, and there is no great difference between the meaning of the silent dance executed by the New Guinea lover before his girl (Guise, 1899) and the posturing of pheasant or warbler. As the female red-necked phalarope or painted snipe displays before the male so does the Minnetaree damsel dance before the man of her choice (James, 1823). In the dance birds and men are kin.

But dancing and posturing have their effect on the bird performing as well as on the spectator—developing energy and releasing it. Dancing cannot be regarded simply as an outlet for the dancers' surplus energy. So far as he is concerned it is a regulative process by which he controls his vitality and maintains his tension at the requisite pitch. The psychical co-ordination attained by mutual posturing or dancing usually more than makes up for the physical vitality expended, and therefore the net outcome is, if we regard the two organisms as a unit, an increase of combined and co-ordinated energy. Negatively, we may consider a good deal of dancing—to use the more inclusive word—as a means of eliminating obstacles to correlation.

There are two types of correlation, which we may call long-term and short-term. The former includes displays and dances of gannets, albatrosses, petrels, divers, grebes, black guillemots, swans, and so forth—birds which even though paired display over a long period and whose posturing is not necessarily a preliminary to the hymeneal rite. Sheld-ducks, though faithful to each other from year to year, will dance their ring dance together continually during winter and early spring. All these activities should be regarded as co-ordination mechanisms fostering mutual psychological and physiological harmony. Under the term 'short-time correlation' may be included those performances which are essentially preliminaries to copulation. Probably the most familiar example is the display of the hen house sparrow; uttering a pleading squeak, *me, me*, and quivering her partially drooped wings she solicits the male, who is ready to respond to a dozen or more invitations (Sterne, 1768).

Whatever the type of the dance, long-term or short-term, erotic or bellicose, solo or social, performed by man, ape or bird there is one generalisation which we can make. In the dance the individual reaches out beyond his isolation and seeks to realise that harmony between himself and the external world without which neither health nor happiness can be achieved and on which the perpetuation of the race depends. It is not surprising that some of the Fathers of the Church said that the angels were always dancing, and that the Gnostics of the second century attributed to Christ the words: 'Whosoever danceth not, knoweth not the way of life.'

CHAPTER XV

ARENA DISPLAYS

The localisation of social displays—Tourneys of the ruff—The great snipe's display—The lek of the blackcock—Importance of territory in arena displays—Performances of the Tetraonidae and other birds—The argus pheasant's display as the extreme form of arena display.

T HE discussion of avian dances has carried us far beyond ceremonial in which only a pair of birds is concerned. In bird display there is an evident tendency to a wider sociability, and we have already noticed that in the communal antics of the red-breasted merganser, sea-pie and black guillemot various stages in the evolution of corporate ceremonies are illustrated. The disposition to display in company may be attributed in large part to the stimulating effect of the presence of other birds and to the interest, curiosity, imitativeness and emotional response aroused in birds by 'goings-on' in their midst. To give only one example, Selous (1901), after studying the autumnal dances of the stone curlew, concluded that 'the birds are stimulated in their dance-antics by each other's présence'. There is, however, another important trend discernible —to localise dances or other antics and establish a rendezvous where birds may display together. This tendency has been referred to in connexion with the sea-pie and has been recorded in species as various as the lapwing (Haviland, 1915; Darling, 1938), sheldrake and bittern (Turner,. 1924). In this chapter we shall consider, for the most part, forms of display which take place at a definite arena frequented by more than one pair of birds. The term 'arena' will be used of the general displaying area; the individual bird's posturing-place, when tended or

kept bare of herbage and litter, will be referred to as his 'court';
the birds which participate in this kind of display will be called
lek birds, using the word commonly employed to describe the
gathering of blackcock. It will be noticed as we proceed that
arena display is much more than the mere localisation of antics
and the concentration of posturing at traditional places. In its
more pronounced forms we shall find it associated with various
peculiarities in the relationship of the sexes, particularly a
tendency to promiscuity.

In Holland the so-called 'hill' frequented by ruffs is usually
on flat ground. The places where the birds posture are entirely
bare of grass, so regular are the suitors' visits and so assiduous
the trampling of their feet. Thus a number of circular courts
a foot or two in diameter mark the scene of the birds' trysts and
tourneys. A wayfarer unacquainted with the strange customs of
the ruff would be greatly puzzled to account for these bare
patches arranged in a pattern in the midst of the flower-studded
meadows. Year after year the ruffs return to the same plot and
attitudinise like feathered marionettes; and to these known
places of assignation come the reeves to choose their mates. So
faithful are the birds to their 'hill' that when a road was made
over one of them they did not desert it. As I passed on my
bicycle they ran from almost underneath the wheels and returned immediately to resume their antics. This strict adherence
to ancestral trysting places is characteristic of many birds with
fixed dancing grounds. The prairie sharp-tailed grouse frequents
the same arena beyond the tribal memory of local Indians
(Farley, 1932), and lesser prairie chickens, even though disturbed, return to their drumming grounds annually (Colvin,
1914). The lone surviving heath hen continued to visit
the ancestral 'booming field' year after year (Gross, 1931).
Beebe (1922), writing of the argus pheasant says: 'An old, old
Dyak chief led me to this arena and told me that his father had

trapped many argus in it, and no one knew when there was not an arena here.' Great birds of paradise resort to the same 'dancing trees' year after year so that the natives have vested interests in their plumes (Goodfellow, 1910), and Chapman (1935) records the constancy of Gould's manakins to certain localities.

The 'fighting' of ruffs is almost as much a misnomer as 'hill' is for the arena where it is staged. There is a great deal of posturing and liveliness, but comparatively few serious combats. The birds rush about excitedly and then stop abruptly in a devotional attitude—heads bent low, legs flexed and feathers ruffled. It is a kind of explosive rigidity, and the ruffs may start off running furiously in another direction or sink slowly quiescent to the ground in a manner suggestive of deflation. At times they leap into the air again and again for a minute on end or dance around one another. They peck at the excrescences on each other's faces and scuffle, feint and spar. Occasionally they fight in earnest, but as a rule no serious injuries are inflicted. During the times of greatest excitement the scene is very animated; the ruffs dart about, crouch and pose, spring and buffet one another with a whirring of wings and thump of bodies, then fly up suddenly in a flock, and presently return. There is something eerie about the unaccountable change of scene. The tenseness of the birds, even more than their frenzied movements, accentuates the impression of sexual stress. It is a strange experience early on a cold spring morning to watch this emotional maelstrom on the darkling polder.

Such is a general impression of the proceedings as observed during the short periods I have spent watching the birds, but Portielje (1931) distinguishes three phases of the display: (1) the ruff scuttles about with head and neck horizontal or bill inclined slightly upwards, tippet raised and wings more or less spread or fluttering, occasionally pecking at the ground; (2) the bird suddenly stops and crouches with the tip of the bill touching the court, ruff expanded, wings half open, tail spread and depressed; (3) he stands with spasmodically shivering feathers and quivering wings—sometimes also vibrating the spread, deflected tail—much as in coition. The sequel to this third attitude may be either that the bird becomes quiescent for a little while or takes an excited flight around, sometimes accompanied by others. Encounters between males take place during phase (1); phase (2) is the solicitous posture assumed before the female.

Selous (1906–7, 1927) showed that the ruffs do not fight among themselves for the ladies nor does the victor in the combat go off with the damsel as his prize. On the contrary, the reeves come to the 'hill' and with calm deliberation choose a male, solicit him and complete their union then and there. Portielje, however, noticed that coition often takes place some little distance from the display-ground. According to him the reeve flies a little way, followed by one or more males, and the female then assumes the crouching, invitatory posture without which the male seems unable to effect coition. Selous says that when the female appears the ruffs flounce about in excitement and then go into the rigid stance; she walks up to the bird of her choice, touches his head or fondles his feathers with her bill and then copulation takes place. Meanwhile the other ruffs remain still. It is extremely unusual for the pairing to be interrupted by them, and the ruffs, when the reeve appears, neither assault her nor each other but allow her to choose quite freely from their statuesque, varicoloured array.

According to Selous (1927):

The common occasion of such fighting as really takes place is not so much the female, as such, as the disturbance caused by any bird flying in to the assembly-ground. Such a one pitching down not quite within its own precincts, disturbs one or other of its nearest neighbours, and each being tenacious of its habitual standing-place, fighting ensues which, in its turn, disturbs others, and so on. This is when any ruff flies in. The reeves, by so

PLATE XV

page 210

RUFFS DISPLAYING AT THEIR 'HILL'

page 210

THE REEVE SELECTS A PARTNER

PLATE XVI

page 220

GREATER PRAIRIE CHICKENS
The left-hand bird is 'booming' and ready to fight.

page 133

LESSER BLACK-BACKED GULLS
Mutual calling.

doing, cause strong sexual excitement, even whilst still in the air. This produces more general excitement and darting about, as a consequence of which a few lightning sparring bouts take place between any two birds who get in each other's way.

This passage elucidates the important point which Selous was acute enough to realise before the 'Territory theory' was formulated by Eliot Howard—that what fighting takes place is in defence of the freehold. It is constituted by the circular depression from which the grass has been worn by the bird's feet.[1] Each of these, as Selous discovered, is owned by one particular ruff. It is any encroachment, or supposed intended encroachment, on this sacred plot which arouses the occupier's anger. This space, too, is often the spot on which the nuptial rite is consummated and in this way subserves the same end as the nest-platform for pairing grebes, divers, cormorants, herons, rooks and other birds. Long ago, indeed, Montagu (1802) noticed that the ruff's altercations were connected with territory. He was shown into a room at Spalding containing about seven dozen males and a dozen females and observed that his intrusion 'drove them from their stands; and compelling some to trespass upon the premises of others, produced many battles'. Thus, according to Mayr's (1935) classification of territory types the ruff, many Tetraonidae and also, as we shall see, Gould's manakin and some birds of paradise belong to his second section, having a mating station not used as a feeding ground and apart from the nesting area. What has happened in the case of the ruff is that the territories of a number of birds have shrunk to the smallest possible size and at the same time drawn as close as possible together.

Before comparing the ruff's display with that of the blackcock let us pause to notice the remarkable performance recorded of

[1] The saplings in the court of the magnificent bird of paradise are frayed so continuously by his beak and feet that they die (Rand, 1940 a).

another Wading bird, the great snipe. So far as can be gathered from the rather incomplete and tantalising observations of various naturalists (Gadamer, 1858; Rohweder, 1891) the proceedings are on these lines: As dusk is falling over the northern marshes the males assemble in favourite arenas, some three to four hundred yards in area, flying with a special slow wing-beat and making a loud *wuff, wuff, wuff,* as they come. Each bird, as soon as he has settled, fluffs his feathers, cocks up his tail and begins to drum:

Here it sits without turning its head; and when the drumming commences it is begun by a whistling note or two; then comes the snapping note with five or six notes in rapid succession, and then a hissing sound, followed by a note resembling the word *shirr* which note becomes deeper as uttered. When the bird commences its note the head is stretched upwards, and is held thus until the snapping commences, after which it is depressed. Whilst producing these notes the bird is in ecstasy, and raises and spreads its tail like a fan, the outer feathers showing in the half-darkness like two white patches. At a short distance the sound of the notes of the different birds at the drumming-place resembles a low continuous chorus, and is by no means unpleasant; for it may be compared with the song of the willow-wren whilst a strong wind is sighing amongst the branches of the forest trees. (Collett,1876.)

The excited snipe may throw their heads backwards, almost upside-down like 'klappering' storks or amorous cormorants whilst singing together (Seebohm, 1885), and at certain times the birds station themselves in a long line, and the song, started at one end, is taken up by bird after bird along the whole file in regular succession, ceasing in the reverse order. As they end their songs they jump on to tussocks and stand with ruffled plumage, drooping wings spread and tails expanded revealing two white areas. This is the attitude which they assume before the female, uttering at the same time their tremulous *shirr*. According to Alphéraky (1906) the females do not emerge from the herbage until about midnight. The males, which have been prancing about in a frenzied way, sparring with one another and performing ludicrous antics, cease for a while and display with raised feathers and tail, but soon renew their 'fighting'. This weird performance continues until sunrise and then, it is said, each female flies off with a male, to return again at night. In Sweden these antics may be seen from May until the end of June when the females begin incubation. Eight to thirty pairs may take part.

Much as the great snipe's display appears to differ from that of the ruff there is the same localisation at a traditional arena, the same formalisation and ecstasy and apparently similar visits by the females to choose mates.

If now, turning to the Tetraonidae, we compare the displays of ruff and blackcock, we find remarkable resemblances in general pattern. Like the ruffs these birds gather at their assembly ground on individual territories before daybreak; like them they posture before the females and spar with one another. The 'scraps' last only a few seconds, and though resounding slaps delivered by the wings echo over the moor all scientific observers agree that, in spite of accounts of birds being found dead on the ground, the fighting seldom results in a fatality. It is, as in the case of the ruff, mainly 'psychological' or 'display' fighting (Selous, 1909-10; Yeates, 1936). For instance, when a blackcock violently attacks a trespasser in normal feeding attitude he immediately changes over from true attack to 'display fighting' if the intruder shows fight instead of running away (Lack, 1939b). The lek is larger than the ruffs' assembly ground and the individual territories are not so clearly defined. One arena measured by Selous contained thirteen courts although only some ten paces by six in area.

Whereas the ruff stands rigid in the presence of the reeve the blackcock runs about the greyhen in such a way as to show all his adornments to best advantage—the near wing drooped and spread, the crimson-combed head held down, the tail tilted revealing the whole of the Cupid's bow. The greyhens come to the lek to be courted, and the advances made to them are much more definite than anything that is to be observed on the assembly-ground of the ruffs. Copulation usually takes place at the stance of the cock, and contrary to the ruffs' complaisant bearing blackcock try to disturb their neighbours in the act; but unquestionably the possession of territories reduces the amount and effectiveness of interference with copulation. Selous considers that the so-called 'war-dance' executed by the blackcock, ruff and other birds has arisen out of the conflict between pugnacity and timidity in the breasts of the duellers, but its true nature is more correctly stated if we say that it is due to the formalisation which has arisen as a safety device to prevent sanguinary, dysgenic fighting. Powerfully armed creatures, such as elephants among mammals and spur-fowl among birds, use their weapons comparatively seldom.

Lack confirms Selous's discovery that clashes occur between blackcock, as between ruffs, when there is threatened or actual infringement of territorial rights. Selous (1927) says: 'It was when one cock was courting a hen, and the progress of the latter over the arena brought him nearer to another that a combat was most imminent.' The following passage makes it clear that whatever the males may do by way of courtship it is the female who makes the choice:

> She now advances slowly into the arena, courted, as she went, by first one and then another male, often by two or three together…still going on, and thus, in spite of some following her, passing gradually from one male to another—for each has his own more especial domain, like the ruffs. At length, however, one bird seemed more to her liking, she paused more frequently, at length stood still, then crouched, and union was effected.

Five main display attitudes have been noted by Selous and defined by Lack: crowing, 'rookooing', display-fighting, circling and crouching. Crowing is often accompanied by leaps into the air and is social rather than intimidatory, apparently having the function of calling the females' attention to the lek. 'Rookooing' is a dove-like utterance, aggressive in intention and uttered with head and neck thrust forward. In circling, the male apparently does not often circumambulate the female but returns after completing a half-circle. He frequently tilts his tail sideways towards her as he passes in front. Sometimes on the appearance of the greyhen he sinks into the crouching attitude which is characteristic of the ruff at such times. In all of these postures the blackcock carries the lyre-shaped tail outspread, the red head-combs distended and the wings drooped.

Not until we have more data available will it be possible to decide on what grounds reeve and greyhen make their choice and to what extent sexual selection is operative, but we have some useful, though incomplete, evidence available. Selous noticed that two ruffs, both outstandingly handsome, were singled out for feminine favours to a remarkable degree, while the majority of the others suffered from enforced celibacy. In three and a half hours of one day there were twelve copulations, and ten of these were with a bird possessing a fine brown ruff. He came to the conclusion that Darwinian sexual selection operated amongst these birds. But the matter is more complicated than this would indicate. Moffat (1924), who studied four captive birds in Dublin, found that one male became dominant and attracted to himself the two reeves. The other made fruitless attempts to dislodge the despot from his court, but it was not until the 'boss' began to moult and lose his sexual ardour that the subordinate bird was at last able to become master and not only occupy the court but also appropriate the two reeves.[1] Unquestionably the situation was highly artificial.

[1] King penguins which moult before breeding choose as mates birds which have completed the moult at approximately the same time (Gillespie, 1932). Such facts show that there is interdependence between physiological condition, display and successful mating.

Instead of the male being a humble suppliant before the female, as Selous observed, he busied himself with rounding up the straying reeves, using his tippet as a brush to drive them back into the territory. The law of dominance had triumphed over all other factors. The problem, however, is further illuminated in some notes made by Turner (1924) during three weeks' observation on Texel Island. She describes one ruff as 'cock of the walk' although he seldom moved from his own court. Twice a day on three successive days, at 11 a.m. and 2 p.m., he was visited by a reeve who settled close beside him. It would appear from this that the reeve may be faithful to her favourite for a period. Turner also refers to a party of ruffs which visited the Norfolk Broads; one of these, a chestnut-brown bird, delighted in disturbing, not only the other ruffs, but shovelers and birds of other species. She says of him, 'He was one of the two original arrivals, and seemed to consider himself cock of the walk. He frequently scurried round the hill, prodding up sleeping ruffs and arousing their ire. Occasionally he met his match in the other first arrival, which had acquired a conspicuous white ruff. These two birds were more advanced in their breeding plumage than the other members of the party.' On Texel I noticed that pugnacity was not confined to the 'hill', for ruffs would chase one another at a feeding pool. These observations show that 'dominance' is an important factor in the ruff's display. It seems that birds which come into breeding condition first become dominant, and possibly their success in attracting and retaining reeves is commensurate and contemporary with the maintenance of their ascendancy over other males. This, in turn, depends on their vigour and virility—indicated both by the grandeur of their frills and the energy of their movements. Vigour and bright adornments are important in 'display-fighting' to cow other males and also to stimulate the females. This situation may account for the accentuation of virility in lek birds; a blackcock copulated fifty-six times in forty-five minutes (Lack, 1939 b), and Gould's manakin eighty-five times in an hour (Chapman, 1935). Unisexual pairing occurs amongst ruffs, and there is also attempted coition with tufts of grass. Ruffs copulate when on passage in England.

This hyper-sexuality is probably related to the autumnal aggressive displays of ruff (Cottam, 1926), ruffed grouse (Bartram, 1754), prairie sharp-tailed grouse (Elliot, 1897), blackcock (Lack, 1939 b), and robin (Lack, 1939 a). No doubt the unseasonable recrudescence of posturing is due to a partial revival of the intense internal state of spring-time which, reaching a lower threshold, arouses intimidatory rather than sexual display, reinforces territorialism and in some species inhibits migration. Bullough and Carrick (1939, 1940) have shown that in the case of the starling there is a correlation between the activity of the gonads and autumnal display. It is highly probable that this is true of other species. Moreover, almost certainly the vigour and intensity of the spring display is directly related to the state of the gonads. Buxton (1941) says of blackgame:

It seems clear that for several weeks the excitement and the fighting of the cocks gradually increases, while the indifference of the hens, feigned or real, gradually diminishes. Then at about the end of April, the peak period is reached and lasts a week or ten days, when the fighting is furious, side-stepping ceases and mating occurs. The decline is rapid: the tattered warriors retire one by one from the lists, and the hens become busy about their nests. Mating may also occur at chosen places on the hillside, to which a successful cock repairs with his harem to spend the day; but it is on the cricket ground in the early mornings that the ladies' affection is gradually won by the appearance and by the prowess of her lord.

While this account may require some qualification in the light of other observers' reports it indicates that the intensity of the performance at leks waxes and wanes, and that its principal significance is, as Selous believed, to provide stimulation for the greyhens. Possibly, so far as the cocks are concerned, there is reciprocal influence between endocrine activity and the display, the one stimulating the other.

Dominance and breeding success are apparently connected in the ruff and other birds which frequent assembly-grounds. If so, early maturity and the appropriation of a court are strong assets for competitors in the marriage market. Ruffs arrive in their breeding quarters before the reeves (although the young of the previous year arrive first according to Naumann), conforming to the practice of other migratory birds in which males stake out claims before the females return. Turner noticed that if an incoming bird found all the courts on the 'hill' occupied he flew away—thereby reducing his chances of pairing; though Selous found some evidence that ruffs would visit each of two neighbouring display-grounds and Portielje observed reeves do so. A bird which is dominant and at a high point in his sexual cycle is most attractive to the female and probably the reeve is able to detect the male so synchronised sexually as to respond most effectively to her advances. Dominance is an indication of this, and as fine plumage and vigour are so closely related to dominance, sexual maturity and virility, it often happens that a particularly handsome bird is chosen.[1] Perhaps the ruffs are able to detect the precise phase of the female, as sometimes she arouses intense excitement and at other times hardly any (Selous, 1927).

As few of the other Tetraonidae have been studied with the patience and perspicuity which Selous devoted to the blackcock, we are, as yet, without the data necessary for a detailed

[1] Unknown factors are also operative in the relative attractiveness of birds for one another. Experiments with a domestic cock and his harem of thirteen hens did not succeed in revealing the grounds of his choice. He had not the same hen as his favourite every day, but on the whole the same hens were in the upper part of the preference list. The attractiveness of the same hen varied with different cocks. (Skard, 1936.)

comparison of the various types of display, but sufficient information is available for it to be possible to form an impression of many of these performances. A short survey of these follows.

The display of the red grouse suggests that just as the ritual of the blackcock is less rigidly conventionalised than that of the ruff so the antics of this bird of the British moors are less stereotyped than those of its near relative. It should be noted that the red grouse, in contrast with the blackcock, is normally monogamous. Only gradually do individuals tend to draw together at definite places where they establish holdings, defend them and challenge rivals. When several males from contiguous territories approach each other and sparring ensues there is a state of affairs approximating to that characteristic of a blackcock lek. There are similarities in the details of the display, too, such as the crowing and leaping into the air performed by both species (Nethersole-Thompson, 1939). The general impression on comparing the habits of the two species is that the blackcock has advanced a long way in the direction in which the red grouse only shows hints of proceeding.

Such evidence as is available shows that there are two phases in the capercaillie's display. Early in the breeding season Hainard and Meylan (1935) watched four cocks calling, parading and counter-marching in an area about fifty yards across, in the presence of two females; but the birds did not appear to pay each other the least attention. Later the cocks are found in *cantons d'accouplement* or copulation territories, and to these the females resort at daybreak, attracted by the rather feeble crowing and louder wing-claps of the males. Later still the cocks continue to defend a territory which does not necessarily coincide with the copulation territory and appears not to be so strictly demarcated. The effect of the system is to permit sexual intercourse to take place undisturbed. The scanty information available suggests that in this, and perhaps some other species with forest courts a certain amount of social display may precede the occupation of the courts.

Turning to the American Tetraonidae an observer of the sage grouse (Simon, 1940) noticed that the cocks usually remained at a distance of from twenty-five to forty feet from each other 'as if each had chosen his own plot of ground', although sometimes from four to ten cocks will gather round groups of females composed of from six to thirty-two birds. The cocks sally forth from time to time with tail feathers spread fan-wise to spar with their nearest rivals, and after a skirmish lasting a few seconds are usually deterred. The fighting consists of wing-beating and pecking. After a bout the birds stand with tails folded and air-sacs deflated side by side, facing in opposite directions and uttering a rapid clucking. Most of the cocks spend their time strutting about without trying to approach the hens. Sometimes aggressive females drive others away. The female invites coition by squatting and the act was not observed except when she did so. Of five cocks surrounding eight hens one copulated with three different hens while the other four cocks did not copulate at all. On rare occasions a soliciting female was ignored by a cock even when she placed herself directly in his way. The inference is that in this species (and possibly other Tetraonidae) there exists a rhythm within the sexual cycle such that the male is not always able to perform the sex act. If this be so, it is clear that one of the functions of the lek is to enable the cocks to sustain social stimulation and find females available when they are able to copulate.

Greater and lesser prairie chickens defend the same territory day by day and the females visit individuals who are at the appropriate stage in the sexual cycle, sometimes two favouring one male (Allen, personal communication). Gross (1928), describing the display of the extinct heath hen, mentions a dominant bird which chased the others in turn. Then the individuals paired off and crouched motionless opposite one another. One would lunge at the other, then they would 'toot' and thereupon the remaining birds started performing. Cocks from different groups would run forward to spar with others, then they would squat and one would call without any interference from its adversary. The females fed nonchalantly in the midst of the excited males. There are parallels between this display and the antics of the prairie sharp-tailed grouse. More than a hundred birds may assemble at an arena. A grouse runs forth and another comes out to meet him. They advance slowly and instead of smiting each other stand *vis-à-vis* in an ecstatic state with eyes closed. The first to come to himself steals away, then the other, finding himself all alone, goes off also. Two by two all the birds take their turn:

The end of each figure seems to be the same. Two birds squat flat on the ground with their beaks almost touching for about twenty minutes, and when they do this they are out of the dances for that day. The dance appears to terminate by some bird, either a late starter or one more vigorous than the rest, being unable to find a partner to respond to his run. Having assured himself of this, he utters a disgusted clucking, and all the grouse fly away at intervals as they complete their term of squatting. (Cameron, 1907.)

Apparently the union of males and females is only temporary. Other observers record lively fights at the display-ground, which is some fifty to one hundred feet across. The grass is worn off and the ground tramped hard and smooth by successive generations of birds.

The partridge, as Aristotle (*H.A.* ix, 8) and Pliny (*Nat. Hist.* x, 51) noticed, shows a tendency to communal display. Various observers (Witherby *et al.* 1941) report seeing large numbers together, and one account describes how pair after pair ran across a ring formed by the other birds. A trend towards communal display in a more or less defined and limited area has also been recorded of the ptarmigan (Millais, 1909). Great bustards display communally and, like ruffs and blackcock, begin before the females are ready to breed and continue for weeks after they have started to incubate. According to Abel Chapman (1893) the large winter flocks divide into parties in the breeding season. In these groups the females considerably outnumber the males, but there are usually two or more cocks in each flock. The males do not appear to have posturing courts, so that the display seems to be of a type rather different from those which we are considering.

African stone pheasants carry out social displays about which little is known. According to Heuglin (Ogilvie-Grant, 1896), during the breeding season the flocks repair to a playground, and while the hens withdraw into cover the males strut and dance, ruffling their neck-feathers, fanning their tails and trailing their wings. The chorus of calls which they utter is answered by neighbouring parties. Males are said to outnumber females in this species.

The male argus pheasant chooses a level spot in the jungle, usually on the summit of a hill or on top of a ridge and clears it of every weed, leaf and twig. The court thus created is about six yards across. Here the display takes place. It consists of three phases; first, a period of excitement during which the tail sweeps the ground while the pheasant fidgets with twigs and pebbles or stamps about the hen; then, when she stands still, he erects his ocellated plumes and confronts her with a great, vertical, concave fan. In the third phase he bows rhythmically to her (Bierens de Haan, 1926a).

The breeding season, according to Beebe (1922), extends over at least six months, and the birds may be heard calling throughout the year except for the two or three months of the moult. The cock's loud *how-how* is uttered at short intervals night after night when he is in his court and may be heard more than a mile away. The pheasants answer each other's calls. The female's

rather less powerful call, *howowoo, how-owoooooo* is repeated
like the male's some ten or a dozen times in succession (Davison,
1896). The birds are heard at sunset, but become more vocal as
night shuts down. Beebe once heard six argus pheasants calling
at once, but considers this very unusual. When he was hiding
at the court of a Bornean argus a young bird appeared; then the
owner came and, after some scrapping, drove it away. Several
display-grounds were found in the neighbourhood, the nearest
barely fifty yards away. Apparently these pheasants are polygynous, though Beebe saw a bird with a mate drive a second
female away. It seems that a pair associate at the court and roost
together for a few days and then the female goes off to lay and
brood. There are obvious similarities between this pattern of
behaviour and the breeding activities of the capercaillie and
peacock (Pitt, 1931). While the pheasants are together the male
ceases calling. His cry may therefore be considered a love-call
such as has been recorded of the heron (Verwey, 1930) and
female red-necked phalarope (Tinbergen, 1935).

What has the argus pheasant, displaying on his court in the
depths of the jungle far from other males, to do with communal
ceremonies? It might seem ridiculous to suggest that in all
probability the argus pheasant's display represents an extreme
form of the lek or arena display were it not that amongst the
game birds themselves we find that lek performances may be
arranged in series from those species which assemble on an arena
with territories only a few feet from each other to birds, such as
the capercaillie, which station themselves on holdings so situated
that the individual birds are a considerable distance apart. The
argus pheasant's court is plainly a sexual territory, defended by
the male, to which the female is attracted by his calls. Davison
(1896) saw a Malayan crested fire-back pheasant chase an argus
out of his clearing. Beebe says that there is assuredly competition amongst the males. It is possible that the timbre of the

male's cry indicates to the female the degree of his maturity, as Beebe found no difficulty in distinguishing the calls of the young from those of adults, and when he heard six calling could differentiate between them. Courts are maintained year after year, but the nomadic young birds' calls come from no fixed point. How courts are grouped is not known. It may be that a vast area of jungle is the 'arena' and that the argus with its courts as far apart as possible represents the opposite extreme to the ruffs' 'hill'. It is probably biologically desirable that such large birds laden with heavy plumes should not display close together in a forest inhabited by predacious animals.

There is a direct relationship between the distance apart of courts and the loudness of birds' notes. The one extreme is represented by the silent ruffs with courts only a foot or two apart, the other by the argus pheasant whose cries are penetrating and frequently repeated. The capercaillie's crow carries only about a hundred yards, but in a well-populated region each bird has a territory contiguous with others (Hainard and Meylan, 1935). Roughly speaking the strength of the call varies inversely as the distance apart of the territories. Lek birds attract attention in a great variety of ways—wing-clapping, stamping, crowing and booming with the aid of air-sacs.

One more display of the arena type may be mentioned here although the exact species referred to is not certain. According to Verrill (1938) certain trumpeters of South America dance only at sunrise and sunset. Like so many birds which frequent favourite dancing-places they clear away all vegetation and make a smooth place for their performance. This practice is, no doubt, related to the fidgeting with twigs, herbage and pebbles which we have seen to be so characteristic of birds in a tense emotional condition. The trumpeters (according to this rather vague account) dance in pairs, by turn, before a vociferous audience, strutting, leaping and whirling, sometimes turning complete somersaults. Verrill is convinced that the Indians copied their dance steps from the trumpeters. He writes:

As the shadows lengthened and the light faded, another and another bird would scurry from the groups of onlookers and join the dancers, until as the jungle grew dim and the birds' forms became indistinct, the entire audience was dancing, curvetting, fluttering in the air like giant moths, and filling the glade with their queer chucklings and sonorous trumpet-like notes. Then, suddenly, and as if at some prearranged signal, the sounds ceased, and with one accord the flock of birds took wing and with a great roar of beating pinions vanished in the forest.

We are told that these dances take place at all seasons of the year, though this may not be strictly accurate as it is based on information acquired from Indians; but in the occurrence of display out of the breeding season, the cleansing of the arena, the frenzied sparring, rapt concentration and corporate flights, the trumpeters' performance is comparable with that of ruffs. Perhaps Darling's (1938) surmise in regard to the previously described dance by grey lag geese may be relevant: 'I think it may be most reasonably explained as a communal display in which the movements of each bird have a stimulative value for the others irrespective of sex and not merely as synchronising the breeding state between male and female.'

When we review the series of ceremonies described in this chapter the conclusion is ineluctable that the advantages of arena or lek displays must be very great. In spite of devoting a large proportion of their time and an immense amount of energy to posturing lek birds appear to be successful species, surmounting the dangers consequent upon habitually frequenting an exposed area and the handicaps of adornments conspicuous to enemies and cumbrous in flight. It is highly probable that not only is sociability in itself stimulating but that the psycho-physiological effects of pugnacious posturing have a beneficial influence on the race.

CHAPTER XVI

ARENA DISPLAYS AND THE SEX RATIO

The display of Gould's manakin—Cock of the rock—Birds of paradise and their relatives—Display of the lyre bird—Convergent evolution in behaviour—Evidence of a differential sex ratio in lek birds—Arena display as an adaptation to meet a preponderance of males—Colonial nesting and the sex ratio.

W E are fortunate in having available detailed observations of another court-frequenting bird very different from those which we have been discussing, for Chapman (1932, 1935) has studied Gould's manakin with as much attention as Selous has the ruff. In the Panamanian jungle I was puzzled by sounds of what I took to be snapping twigs, suggestive of the presence of a puma, tapir or other large beast. It was not until Dr Chapman told me that these elusive crepitations which had beguiled me for several days were made by Gould's manakin that I learned to connect them with the small yellow and black birds which I had seen tripping excitedly from branch to branch.

On the forest floor the cock manakin makes an elliptical court, some thirty inches by twenty, clearing away every leaf or twig which falls upon it and leaving bare the hard earth as if it had been swept by a broom. It is oddly like the court of the ruff, although amongst surroundings as utterly different as could be conceived. These courts are in groups varying in distance apart from twelve to two hundred feet, though the most usual distance by which they are separated appears to be about thirty feet. A group of courts may be considered an 'arena' in the sense in which we are using the term. Groups tend to occupy the same sites year after year, but the courts may differ in their relationship to each other. The terrain favoured is forest diversified by young saplings.

During the daylight hours of the breeding season the manakin frequents his court and its surroundings. He flies briskly away to feed every two or three hours but returns within a few minutes, occasionally carrying a berry in his bill. At intervals, when he notices a female in the neighbourhood, or the spirit moves him, he begins his display. Flitting in a tense way from perch to perch, with the long feathers of his throat erected and thrown forward beyond his bill to form a projecting 'beard', he calls *pee-you* with increasing volume and rapidity. Suddenly, lifting his wings, he makes a snapping *whirr-r-r* with them and cries *chee-pooh*. He has now reached the borders of his court and enters with a loud *snap* which may be heard for a distance of three hundred yards; *snap-snap*, he springs from side to side, alighting on the saplings around the court about a foot from the ground. The manakin may now settle in the court and clean it, if necessary; then he hops from root to root within its precincts making a low *snip-snip*, always managing to alight facing the spot from which he jumped. Sometimes a bird will press the tip of his beak into an upright stick and flutter like a moored dirigible, or even alight and poise in the court head downwards with vibrating wings. At other times he will stand with head to one side or pointed upwards in a rigid gaze-pose lasting from several seconds to a minute or longer.

These odd, acrobatic activities have as their function the attraction of the female. Her leaf-green colour renders her extremely difficult for human eyes to detect, but male manakins manage to perceive her when she is still a long way off. Immediately all is excitement—just as when a reeve visits the ruff's arena. Each court resounds with snaps as the cocks animatedly begin to display. The female proceeds deliberately to the court of her choice, but the favoured manakin does not press his suit directly nor endeavour to force himself in any way upon her. By jumping back and forth in the court he issues an invitation to the dance similar to that of the cock wheatear (Lloyd, 1933). If she chooses to accept, she joins in and the two birds leap rhythmically across the court, passing each other in mid-air—a duet even more impressive than the cedar waxwings' exchange of berries or the red-throated divers' 'penguin' dance on the lochan. Then the couple disappear, doubtless to consummate their union.

Other species have related customs. A naturalist in Nicaragua watched two male manakins on a branch leaping about two feet into the air and alighting on exactly the same spot. Their procedure was as regular as clockwork. One bird jumped as the other alighted, each bird accompanying itself to the tune *to-le-do*—*to* as he crouched, *le* in the air, and *do* as he perched. They continued this without intermission for more than a minute (Nutting, 1883). The red-headed manakin, which I have watched performing in the jungle, postures before the female like a ballet dancer on tip-toe or jumps sideways back and forth with occasional half-whirling pivots. The birds display either in pairs or parties and habitually resort to a favoured place for the performance (Chapman, 1929).

Manakins, in spite of their rivalry for the females and apart from the dominant-submissive relationship shortly to be referred to, are on excellent terms with one another. As a rule court rights are strictly respected in the manakin world and there is seldom active hostility, but when this does occur it is due to interference with, or intrusion upon, another bird's court. When quarrels develop the court is neglected and displaying restricted and hampered. It is impossible to fight and woo effectively at the same time. Courtship is primary and antagonism secondary in manakin affairs, and in spite of some appearances to the contrary the general rule applicable to all leks is that it is interference, or the threat of interference, with the court on which the individual bird displays which arouses antagonism rather than the mere presence of potential rivals.

Chapman was able to discover the exact limits of a manakin's territory. The bird on which he experimented would respond to a stuffed female placed on a branch by first snapping in court and then approaching the mounted specimen. Of any two birds which came into competition with each other one would tend to be dominant, the other submissive. When the dummy was placed midway between two courts, both birds wooed, but the dominant manakin claimed her. But when the specimen was moved towards the court of the submissive male a point was reached beyond which the rival refused to go. Here we have proof of mutually recognised territorial boundaries essential to the functioning of the whole pattern of behaviour, and also insight into the nature of territory. It is compressible up to a point, but the nearer the trespasser gets to its centre the more vehement is the resistance of the proprietor, and the less stomach the trespasser has for continuing the contest.

The court and its associated territory thus serve the two-fold purpose of providing a known rendezvous where the female may meet the male, and of limiting the scale and range of fighting by reducing it, when it occurs, to disputes for territory. Thus duelling or general mêlées arising out of the desire for mates are avoided. Nature side-tracks the bellicosity which sexual rivalry always involves by the device of concentrating the bird's energy upon the defence of a plot of land.

In one respect there is a marked difference between manakins and ruffs. The latter, as we have seen, are amongst the most silent of all birds. No sound is uttered at the 'hill', though I have noticed when close to feeding birds that they make a scraping or croaking noise; the reeve also has a low, guttural note. On the other hand, manakins continually make their loud snapping, and in some species the quills have become greatly modified to enable this to be done so effectively that this nuptial invitation can be heard hundreds of yards away. The loud *snap* is of great importance as a signal to guide the questing female to the male awaiting her in the depths of the shadowy forest. The crowing and leaping of blackcock and the dancing and corporate flights of ruffs serve the same purpose of enabling the females to find the males without delay. The blackcock's crow has weak carrying power compared with the song of a typical Passerine bird, and this is probably to be attributed to the small area of his territory and its contiguity with other territories. It is natural, therefore, that the ruff should not use his voice at all on the tilting grounds as he occupies a territory reduced to the smallest possible dimensions. But the birds remedy their vocal deficiencies by conspicuous corporate flights and lively antics. What reeve on the polder is likely to remain unaware of what is toward amongst the ruffs? No loud call is necessary to warn off trespassers, for ruffs spend their time on each other's doorsteps. *Per contra* we thus have confirmation of the function of song to warn off intruders and attract a mate.

A larger relative of the manakin, the beautiful cock of the rock, performs communal antics which have yet to be adequately studied. The main details were recorded by Schomburgk (1841) more than a century ago. He speaks of the arena as a space four or five yards in diameter carefully cleared of every blade of herbage, but Frost (1910), who saw a large number of 'dancing-grounds', refers to them as two and a half to three feet across and says they are invariably under trees where the birds feed and often near running water. The main dance is between 8 and 10 a.m. The Indians believe that the birds dance at no other time, but Frost caught some in the afternoon. Arenas are frequented for at least some weeks, but if the birds are disturbed they will start another elsewhere. Schomburgk's account tells of a bright orange male which danced, stretching his wings, raising his head and expanding his tail like a fan. When he was tired of strutting proudly about another took his place. This posturing in rotation has been noticed in connexion with albatrosses as well as the prairie sharp-tailed grouse (Bent, 1932), but it does not agree with Frost's observations—which appear to be more reliable. He watched thirty or forty dances and gives the following particulars about them: They are started by an old cock who squawks from a fallen or sloping branch by the arena and jumps back and forth from his perch to the ground, giving a flick of his open wings and tail as he touches the bare earth. One after another the birds cease feeding overhead, descend and stand around watching. Then one or another joins in, perching opposite the first and taking turns in the jump. More participate until there may be seven or eight performers hopping back and forth.

According to Verrill (1938) one, two or three birds dance together before spectators. So far as one can gather from his rather vague and doubtful account, the phases of the performance approximate to those recorded of the ruff and blackcock. There is a jerky circular parade. Every few moments two face each other and bow their heads until their beaks touch the ground. A third figure is similar to the sparring of other lek birds. They rush at each other with croaking cries, spring into the air and hover for a moment or two, feinting to strike; but all this show of battle comes to nothing. Every now and then one of the displaying males leaves the arena and flies off with a female. But Frost states that he saw no females at the dance. All the birds which he captured supposing them to be females turned out to be immature males. Chapman (1935) shows how easy it is to be mistaken as to the sex of manakins. Seeing four bearded manakins in Trinidad displaying noisily and vigorously he shot the only specimen which was in female plumage—to discover on dissection that it was an immature male. Birds of paradise in immature plumage are also seen at the *sacaleli*. The dances of the cock of the rock continue even when the females have eggs or young, as is the case with several other species practising lek ceremonies.

Until we have more satisfactory data it would be folly to attempt an interpretation of these ceremonies, but we may note that the to-and-fro jumping recorded in the most reliable of these descriptions is reminiscent of Gould's manakin. Frost tells us that the birds are extremely vicious to one another when caged. This suggests that as at the leks of other species the emotional basis of the gatherings is rivalry.

The so-called lek ceremonies of humming-birds will be mentioned later when we consider aerial displays.

When Wallace (1869) compared the movements of the king bird of paradise with those of the manakin he was wiser than he knew, for there are similarities in the modes of display:

It frequents the lower·trees of the dense forests, and is very active, flying strongly with a whirring sound, and continually hopping from branch to branch. It eats hard stone-bearing fruits as large as a gooseberry, and often flutters its wings after the manner of the South American manakins, at which time it elevates and expands the beautiful fans with which its breast is adorned.

Rand (1938) has shown that probably the male occupies a territory of one or two trees and he surmises that the female comes there to mate. Perhaps the habits of the species may be regarded as in some respects intermediate between the paradise birds which display socially in trees and those which keep courts like Gould's manakin.

The display of the lesser bird of paradise also has analogies with that of Gould's manakin and phases akin to the ruff's. A score or more gather on a favourite tree and work themselves into an ecstatic condition, quivering their wings and calling

PLATE XVII

page 236

MALE SATIN BOWER BIRD ADDING
A LEAF TO HIS BOWER

page 20

WATER RAIL REMOVING HER EGG

PLATE XVIII

page 233

GREAT BIRD OF PARADISE
IN DISPLAY

page 233

COUNT SALVADORI'S BIRD OF PARADISE
IN DISPLAY

rapidly *walk, walk, walk*. Then the wings are suddenly raised, the tail depressed, and with a slight rustle the golden, diaphanous side-plumes are erected, forming an arched cascade over the back. In this lovely pose the bird holds himself motionless except for the quivering wings and tremulous plumes. Then, together with the other members of the assembly, he hops wildly along the bough with plumes uplifted emitting harsh cries, *ca, ca, ca*. After these frenzied activities the bird stands with arched back in an ecstasy, rubbing his beak against the bough—a not uncommon feature in sexual displays (Ogilvie-Grant, 1914). Not only do the three phases resemble those of the ruff but there is similarity in details such as the depressed tail and nervous fidgeting. In the case of the ruff this consists of aimless pecking at the ground and in the bird of paradise needless cleaning of the bill.

The antics of the great bird of paradise (and the closely related subspecies named after Count Salvadori) conform to the same pattern—at one time frenzied, now with plumes a-quiver, then static. Wallace (1869) speaks of a dozen or twenty males joining together in *sacaleli* or 'dancing-parties' together. Goodfellow (1910) saw sixty or seventy feeding in company. They assemble in lofty trees—some of them even before they have donned their full nuptial plumage[1]—and pass from branch to branch with much rapidity and

in great excitement, so that the whole tree is filled with waving plumes in every variety of attitude and motion. The bird itself is nearly as large as a crow, and is of a rich coffee brown colour....At the time of its ecstasy, however, the wings are raised vertically over the back, the head is bent down and stretched out, and the long plumes are raised and expanded till they form two magnificent golden fans striped with deep red at the base, fading off

---

[1] The significance of communal display amongst immature oyster-catchers, cocks of the rock, trumpeters, manakins and birds of paradise deserves investigation. According to Lack (1940b) the Galapagos finches display vigorously and often breed in juvenal plumage.

into the pale brown tint of the finely divided and softly waving points. The whole bird is then overshadowed by them, the crouching body, yellow head, and emerald green throat forming but the foundation and setting for the golden glory which waves above. (Wallace, 1869.)

The main dance is at 7 a.m., though there is a minor 'hop' in the afternoon. These birds are believed to be polygynous and Goodfellow states that certainly this is true of *Paradisea apoda* and other members of the genus. Possibly they are promiscuous.

Count Raggi's bird of paradise and other species also congregate in trees to display, but the six-wired bird of paradise performs in a way which bears a close resemblance to the antics of manakins and cocks of the rock. A plot of ground on a ridge some four yards in diameter is cleared, and from low branches bare of leaves the birds hop to the ground and back again, displaying their plumes (Simson, 1907). The magnificent bird of paradise not only picks leaves and twigs from the ground but even tears leaves from overhanging shrubs to allow the light to shine on him in his court. Experiments showed that a male attacked a mounted male once but ignored a mounted female (Rand, 1940 a).

Those less gaudy relatives of the birds of paradise, the magpie and the jay, participate in revels which resemble crude versions of the New Guinea birds' display. In late winter from six to two hundred magpies may be seen excitedly hopping amongst the branches of trees or hedges, raising and fanning their tails and simultaneously erecting their head-feathers. There is a good deal of flying back and forth and general commotion. The males sometimes hover like kestrels a foot or so from the ground in front of the females. Raspail (1901) claimed that these gatherings occur when one member of a pair has been killed, and it is recorded that they are held annually at the same place (Stubbs, 1910). Jays congregate in spring, posture and pursue one another. During their ceremonies a soft, crooning song is heard, and, according to my observations, a buzzard-like mew. Several will fly in succession with slow butterfly-flight like the display-flight of the oystercatcher. To what extent these ceremonies have a nuptial significance is not known. Small flocks of displaying magpies may be seen at all seasons, and pairs of these birds remain together throughout the year. Possibly it will be found that in such social gatherings the birds stimulate each other and thus facilitate the reproductive processes. They may well have a function comparable with that of the communal displays of gulls.

Ceremonies of a somewhat similar kind have been reported of the Australian white-backed magpie. The birds are said to mate for life, but in mid-April several pairs with their grown young assemble on neutral ground. The ceremony opens with a noisy chorus and all the birds seem to be on good terms with one another. This state of affairs lasts several days and then the old birds withdraw while the young pair up. When the parents return they drive their offspring away (Hall, 1909).

Although bower birds have attracted much interest since their remarkable habits were first commented on more than a century ago, we still lack detailed, scientific life histories of the various species. Indeed, descriptions of them are often not only confusing but contradictory. Writing of the regent bird Phillips (1911) says:

As the nesting season draws near, the adult females withdraw from the General Bower, scatter, and each makes a private bower for herself, somewhat of the shape of a horse-shoe magnet....In these Nuptial Bowers, each by herself, the females, who are understood to outnumber the males, use all their arts to attract a male. For a while the two birds are all in all to one another in and around the Nuptial Bower; the female commences to build, the Nuptial Bower—or what remains of it—is scattered to the winds, the spot is deserted, and the male goes off after female Number Two in another Nuptial Bower elsewhere, leaving the first to finish her nest by herself.

This would suggest a state of affairs somewhat resembling that which we shall shortly discuss in connexion with Wagler's oropendola, but other accounts of the behaviour of this species disagree on important points (Söderberg, 1929). Marshall (1934), speaking of the satin bower bird, says there is no evidence that coition takes place in the vicinity of the bowers although the male displays to the female there. It may be mentioned that the satin bower bird's practice of painting sticks with berries or charcoal for paint and fibrous material as brush or 'stopper' qualifies it to enter the small group of avian tool-users (Chisholm, 1934; Gannon, 1939). Amongst these are the tailorbird (Hume, 1890), the little spiderhunter, which employs its bill as a needle and spider's lines as thread (Shelford, 1916), and the Galapagos finch (*Cactospiza pallida*), which uses a spine to poke insects out of crevices (Lack, 1940b; unpublished).

In regard to the lyre bird, despite numerous more or less popular accounts, there are still many gaps in our knowledge. About the end of April the males prepare new courts and repair the old. These are kept bare and are usually raised a few inches. The birds elevate their tails and utter a remarkable repertoire of calls and an elaborate song at one court after another, using as many as eight or more in a morning (Tregallas, 1931). One bird which was studied for several months possessed more than twenty of these mounds. The observations showed that both male and female occupied territories, but there was a neutral area in between where the birds associated. Campbell (1941) noticed that the male's song could be heard clearly at the nest 250 yards away. Copulation took place at a mound after a chase of 1150 yards from one mound to another. It was noticeable that when the bird tried to display elsewhere than on his courts the ferns interfered with the erection of his plumes. In the area investigated there were two main territories and at least six birds, but some of these were immature. Indeed, even whilst the male was courting the female she was accompanied by a young bird. It should be noted that any difficulty which the female might have in finding a male at any one of a score

of courts is offset by his loud and persistent song, equivalent to
that of an ordinary territorial song-bird and more continuous
than the utterance of any other court-keeping bird.

The maintenance of a court is incompatible with any considerable measure of attention to the domesticities, and thus we
find that amongst these birds the males take little or no part in
the building of the nest, brooding the eggs or tending the
young. The nest-building impulses are diverted into the upkeep
of the display area, and energy which in other birds is used up
in family chores is diverted into posturing. Just as in primitive
societies marriage customs and the system of land tenure are
closely bound up with one another, so amongst birds the behaviour-patterns are very subtly interconnected.

These examples of court display effectively illustrate convergent evolution in behaviour. Groups as widely separated as
the Game, Wading and Passerine birds participate in ceremonial
with strikingly similar pattern and figures. Moreover, amongst
them assiduous display, extravagant development of adornments
and sexual selection seem to be the rule. Very many of the
birds which we have considered clean their courts with great
care and pose ecstatically in a rigid stance as well as display in an
excited fashion. It is reasonable to suspect that, as with convergent evolution of structure, so this parallelism in behaviour
evolved as a solution to a common problem. What can this
be? What difficulty can there be which besets the ruff on his
polder, blackcock on the moor, manakin and bird of paradise
in the jungle? I suggest that the ceremonies of these birds may
be adaptations to meet the problems involved in a differential
sex ratio—or possibly, what in practice amounts to the same
thing, a differential ratio between the numbers of each sex con-
temporaneously in an effective breeding condition. This suggestion is put forward somewhat tentatively, as our information
on the subject of the relative numbers of the sexes in lek birds

is meagre and not always reliable, but some of the data are worth reviewing in order that future workers may be stimulated to study a field in which physiological and psychological factors are closely interrelated.

Most observers agree with Gordon (1920) that blackcock are more numerous than greyhens, but the relative numbers at the lek vary and his statement that the males present outnumber the females by ten to one gives a misleading impression. Observations at the lek, necessarily somewhat inconclusive, are supported by the evidence of the Swedish *Tidskrift för Jägare* (Lloyd, 1867) that the broods of blackgame contain more males than females. This also holds good of the capercaillie (Lloyd, 1867), which other observers besides Hainard and Meylan have noticed showing a tendency to hold tournaments. These two naturalists found twice as many males as females in the area studied. At the *lek-ställe* of willow grouse the males outnumber the females. This has been attributed to the greater dangers to which the breeding females are subjected (Darwin, 1871), though it has never been proved that a brightly decorated, frequently displaying cock bird in the open is safer than a protectively coloured female crouching on her nest. The same explanation has been put forward to account for the great excess of males over females in the extinct heath hen. At a strutting ground of the sage grouse Simon (1940) noticed about 300 cocks and 80 hens. Those who have studied the ruff believe that there is a deficiency of females (Weir, 1871). The maximum number of reeves visiting the 'hill' watched by Selous (1906–7) was fourteen or fifteen as against twenty-two or twenty-three ruffs, and out of this number of males only eight at the most were seen to copulate. A preponderance of males over females is strongly suspected in cocks of the rock and birds of paradise. There are great disparities in the sex ratio amongst hummingbirds. Salvin was convinced that males preponderate in most of the species; in one year he collected 204 specimens of ten species and of these 166 were males and thirty-eight females (Darwin, 1871). An American student of the humming-birds of California notes that in spring males are in the great majority, a state of affairs only partially to be accounted for by the more retiring habits of the females (Woods, 1927). On the other hand, Mayr (1939) says the females outnumber the males in many genera of the Trochilidae, and Nicholson (1931) found an apparent sex ratio in the Guiana king humming-bird of eight males to thirty females.

Let us suppose that for some reason the sex ratio in a species holding territory around its breeding site is upset and a situation arises in which there are more potent males than females. Wait they never so patiently, proclaim they their territorial conquests never so loudly and assiduously, some of the males never could obtain a partner. If, in consequence, the territory convention broke down, as it would tend to do, and the males roamed about in search of mates, the whole behaviour-pattern would be disrupted and the outcome would be dysgenic. It is significant that when male manakins leave their courts to pursue the females in the tree-tops 'the result is confusing and apparently the ends of neither sex are served' (Chapman, 1935). Formal places of assignation offer better prospects for the male and female to meet and synchronise their sexual rhythms than would fortuitous encounters of nomadic birds. In monogamous species sexual jealousy, issuing in combats for mates, would interfere with breeding. Polygyny, with the maintenance of the breeding territory, would, and actually does, solve the problem in some species, but the lek birds have discovered another solution.

Territories remain display grounds, but may shrink and converge on one another to form a system of courts within visual or auditory range. The permanence of the rendezvous and the conspicuousness of the performers make it easy for the females to seek stimulation and find partners when they are at or approaching their peak of desire. None of the females, so valuable to the race, need go un-inseminated nor absent herself long from the eggs which she alone incubates. Some lek birds are highly sexed and the females remain ardent for a long period. It might be expected that such birds brought close together on an arena would fight furiously, but the disastrous effects of fierce antagonism are avoided by associating bellicosity with the defence of territory and by stylising the fighting to the point of being practically harmless. Social posturing, it may be noted, seems to have a cumulative effect, so that birds which perform tend to do so more and more. We may expect that the physiologists will one day throw light on this.

It is not suggested that a lek always tends to be formed when the sex ratio favours males, for this is clearly not the case. Many observers have noted a preponderance of males in starlings (Bullough, in the press) although these birds show no inclination to form leks, but in such instances other factors are operative. For example, the fact that females breed in their first year and males in their second may appreciably restore the balance between the sexes, although Kluijver (1935) finds a predominance of males even in breeding populations. The starling (Freitag, 1936) and other usually monogamous birds such as the snow bunting (Tinbergen, 1939a) will take mate after mate when the female reaches the non-copulatory stage. It is significant, so far as a tendency to lek formation is concerned, that female starlings choose their mates (Morley, 1941). Various species solve their breeding problems in different ways, but, as Mayr (1939) has pointed out, unequal sex ratios are nearly always found to be correlated with peculiarities in the life history and breeding behaviour of the birds. Polyandry is not often resorted to and appears to be confined mainly to those species, such as button quails (Seth-Smith, 1907) and painted snipe (Pitman, 1912; Baker, 1935), in which the female is the more brightly coloured of the pair and the rôles of the sexes are, in many ways, reversed. But there is no convincing evidence of polyandry in the phalaropes (Tinbergen, 1935). Polygyny is much commoner than polyandry and may even occur in species with a surplus of males (Nice, 1937). Probably both territory and dominance drives militate against polyandry as a solution of sex-ratio problems, for polyandry is only possible when the characteristic male impulses of sexual jealousy, superiorism and proprietorship are subordinated or, perhaps, formalised into conventional or ritual activities.

Darwin (1871) compared the extravagant adornments of Gallinaceous birds with the long golden sword of a fish now common in aquaria—the swordtail *Xiphophorus helleri*. In this viviparous species there is an excess of males (Noble, 1938). It would be precarious to found any theory on the correlation of exaggerated decorative features and an unequal sex ratio in some species of fish and birds, but the extent to which such a correlation exists—or does not exist—is worthy of examination. If it could be shown to hold, exceedingly interesting possibilities would suggest themselves in regard to the puzzling question, the origin of a differential sex ratio.

Let us now consider what happens in some of the species in which the females are known to outnumber the males. Chapman (1928) established the fact that there are at least six times as many females as males in a colony of Wagler's oropendola. These birds nest in groups, but although they begin operations with astonishing approximation to the same date each year, the females, who do all the nest-building, commence at different times. Thus they are ready for insemination at intervals and in sequence. Whilst the female is busy with the nest—a beautiful, woven structure—she pays no attention to the ardent wooing of the male. He leans down, ogling her with his blue eyes, raising the feathers of his crest and lower back and twitching his tail, the while agitatedly spluttering, cackling, wheezing and gurgling. So he continues hour after hour, day after day, all to no avail. It never enters his head to assist her with the nest, or her's to pay him the slightest attention. But when the nest has been woven and all things have been made ready she responds to his advances, caressing the feathers on his head with her bill and spending several days in amorous dalliance. It is during this period that the eggs are laid. But when the time for incubation arrives all is changed again. She composes herself to brood the eggs forgetful of him, and he goes off to seek another temporary mate, who, having by this time finished her nest, is ready to receive him. The oropendolas thus solve their problem in a way which avoids quarrels between rival males, assures the sexual satisfaction of the birds and the successful propagation of the species. Chapman calls it 'limited monogamy'. Instead of the males fighting for territories the females struggle to acquire and defend a suitable site for the long, pendant nest. Visitors from other colonies are received with tolerance. If the oropendolas' nesting tree is blown down they colonise another tree close at hand. Thus by adhering to a traditional nesting- and trysting-place the meeting of synchronised birds is facilitated.

The habits of the oropendola may be compared with those of the crimson-crowned bishop bird as described by Lack (1935). Of five males which were studied, one had one female, two had two females, one had three, and one three or four. The sex ratio appears to be one male to four females (Moreau and Moreau, 1938). The male courts one female at a time, builds a nest for her and then is ready for a new female. He has a strong sense of territory and patrols the boundary, tilting at other birds, but avoiding trespass as if there were an invisible net between the domains. For this species territory has no food value, so that as Lack points out its importance probably lies in the isolation provided for the cock and the consequent ease with which the female can locate him. What the oropendola achieves by transferring territorial jealousy to the female (in so far as there is rivalry for nesting sites) the male bishop bird attains by emphasising it. When the sex ratio in birds is in favour of the males, they may establish a conspicuous social display while retaining a strong territorial sense, or, in some tropical species, they may co-operate with mated birds; when males are deficient in numbers, either the females may establish conspicuous social nesting colonies or the males may make themselves particularly conspicuous by assiduously demonstrating on their frontiers.

The American boat-tailed grackle is another species in which females are more numerous than males. McIlhenny (1937) worked out the sex ratio with great exactitude and found by examining nestlings and trapped birds that there are rather more than twice as many females as males. This handsome and conspicuous bird, clothed in metallic blue-black and purple plumage shot with green, inhabits tidal marshes in Florida, Alabama, Mississippi, Texas and Louisiana. The cocks gather in parties on the ground or on the tops of coarse grass or low bushes, perched perfectly still, facing one another. The pose is statuesque, beak stretched upwards, tail depressed and plumage compressed tightly over the body. Excitement at such times makes the usually hazel-coloured eye gleam golden. Now and then when one takes flight the others follow with rattling primary quills and repeated squeals and throaty gurgles; but they soon alight again and take up the same rigid posture.[1]

[1] These stiff poses in display are of considerable interest. After spreading his wings and uttering a squeaky whistle the jet black Nicaraguan grackle compresses his plumage and points his bill skywards (Belt, 1874). Hudson's tyrant-bird, an intensely black-plumaged species, displays elaborately before the drab female. Every now and then he begins revolving round his elevated perch like a moth fluttering about a candle flame. When first he flies off he suddenly reveals the hitherto concealed white wing bars, flashing brightly

When a female feels so inclined she flies to one of these bachelor parties. Immediately they see her the males begin squeaking, fluffing out their plumage, spreading their tails and rattling their quills, each bird making the most of his appearance. Then they rise in a body and follow her. If she does not find one of them sufficiently attractive she outflies them all and goes on to other groups until she discovers a suitable match. She makes her choice known by flying in front of, and close to, the favoured bird, edging him to one side out of the flock. The others relinquish the chase and the pair alight and consummate their union.

Thus it appears that boat-tailed grackles are promiscuous in their relations, for the female leaves her partner after coition and he makes no attempt to follow or associate with her further. As a rule the males take no interest in the nests except that they will occasionally eat the young, as they eat the chicks of other species. But a few of the cocks constitute themselves in some sense guardians of nesting colonies, and although apparently

as the wings open and shut, but as he performs his gyratory dance the black and white on the quills fuse and it seems as if a grey mist envelopes the bird's black body. The wings hum and a series of clicks is heard. As suddenly as the antics began they come to an end, and the bird drops to his perch to remain 'as stiff and motionless as a bird carved out of jet' (Hudson, 1892). Millais (1913) saw a satin bower bird in the London Zoo pick the only daisy in his pen and hold it for twenty minutes in the face of the female. She also remained stationary. When emotionally excited, cranes will throw objects into the air, catch them, and hold a rigid poise for a time. The connexion between static posturing and dark plumage deserves investigation. The colour black has a strong emotional valency and is common as a psychological weapon. Also, to stand perfectly still is one of the most surprising things such a volatile creature as a bird can do. Not motion itself but contrasts in it most readily attract the attention and are therefore supremely effective in display. Such birds as the ruff, the manakin and the Paradiseidae will pass in the space of a few moments from frenzied motion to complete immobility. The psychological mechanism involved may be comparable to that of the 'injury-feigning' trick.

they have no direct sexual relationships with the females they become very agitated when any danger threatens the nests. As with the oropendolas, so with the grackles; there are no individual territories (with the exception just mentioned) and no fighting or sparring amongst the males.

I interpret these facts as indicating that territory amongst birds is primarily the centre of sexual activity rather than the feeding ground, although the 'feeding-ground territory' is a well-known and definite phenomenon. Night herons, for example, will quickly establish feeding preserves away from their nesting territories (Lorenz, 1938). Rather than speak of territory as the 'mating area', as some do, I prefer to regard it as the centre of sexual activity, for some birds such as hawfinches, cedar waxwings and various auks, ducks and gulls form pairs before they have occupied territory. Lack (1939a) showed in his study of that most territorial of birds, the robin, territory's chief importance to consist in facilitating pairing-up.

With many provisos and reservations we may tentatively suggest that where the sex ratio favours males they tend to display colonially; when it favours females they tend to nest colonially. Of course communal display or colonial nesting may, in some species, be due to other factors. Amongst boat-tailed grackles the social habits of the males make it easy for the females not only to find them but to select a partner at the requisite stage in the sexual cycle. Colonial nesting, in conjunction with the conspicuous habits of the males, prevents the diminution of breeding efficiency which would result if the females could not readily find partners when they wanted them.

Climatic and topographical factors also have an important influence on nesting. Colonial breeding is only possible in circumstances when there is a sufficient degree of protection from marauders. The scarcity of suitable sites forces many sea-birds to nest in company. Temporary monogamy, the device

by which the oropendola solves the problem of the deficiency
of males, avails for species nesting in the tropics with an abundant supply of food obtainable over a lengthy period. Staggered
nesting would not work where a species is dependent on a
strictly seasonal food supply. Promiscuity is the solution
adopted by the boat-tailed grackle, in which species females are
twice as numerous as males, but other birds with a surplus of
females, such as the American tricoloured redwing are polygynous and occasionally promiscuous. This is 'perhaps the
most colonial of all Passerine birds', nesting in colonies up to
200,000 strong. It is social by preference, not by ecological
necessity. The males hold territories some six feet square and
attract the females by erecting their epaulets, spreading their
tails and fluttering down the cat-tails as an invitation to the
prospective mate to follow and nest. But territory, fighting,
threat display and territorial 'advertising' song are feebly developed, so that Lack and Emlen (1939) think they may be disappearing in the evolution of the species. Probably here we
have a stage through which the boat-tailed grackle has passed
to promiscuity and the disappearance of nearly all territorial
behaviour.

Although we must await the results of further research before
we can speak positively of the bearing of the sex ratio on bird
ceremonies and the appropriation of territory yet this short
survey has shown how closely connected and complexly related
are physiological, psychological and ecological factors in the life
of birds. Just as the anthropologist traces patterns of culture in
human society, so the ornithologist finds that bird behaviour
may be best interpreted as a subtly connected interdependent
system or complex. Territory, display, gland-secretion, sex
ratio, plumage, climate, food-supply, size of brood, migration
and so forth, make up a pattern of behaviour in which we can
discern, but not yet confidently follow, the intersecting strands.

CHAPTER XVII

TERRITORY, SONG AND SONG-FLIGHT

Definitions of territory—Territory not primarily a feeding area—Sexual jealousy the basis of territorialism—The function of song—Correlation between song-flight and territory—Display-flights—Pleasure and sexual stimulation associated with display-flights.

THE demarcation and defence of territory by birds was recorded by Aristotle (*H.A.* IX, 32) and Pliny (*Nat. Hist.* X, 5). It was more precisely formulated by Bernard Altum in 1868 and brought independently into the field of scientific discussion by Howard (1907–14, 1920). As set forth by him the theory may be concisely stated thus: The guarding of a specific area (usually around the nest) is of value to birds because it distributes them regularly, thereby reducing the chances of birds remaining unmated; it also serves to strengthen the bond of union between the pair and guarantees the family's food-supply, especially at the critical time when the chicks are newly hatched. The controversy with regard to this theory and the criticisms elicited by Howard's views (Lack and Lack, 1933) have resulted in a clarification of the problems involved and a great deal of *ad hoc* research. On the one hand it has become evident that counsel is darkened if the term 'territory' is used without definition; on the other, as the term has psychological as well as ecological implications, it is very difficult to formulate a definition valid for the wide range of territory-holding animals; amongst others ants (Elton, 1932), bees, both breeding and feeding territories (Kalmus, 1941), fish (Noble, 1938), lizards (Noble and Bradley, 1933; Evans, 1938), birds, seals (Alverdes, 1935), red deer (Darling, 1937) and monkeys (Carpenter, 1934). It will be of advantage, therefore, to seek out the psychological drive which is responsible for the establishment and maintenance of territorial rights amongst birds. Is the guarding of territory primarily due to pugnacity, the desire to protect feeding-grounds, sheer possessiveness or some other motive? If we can detect the primary drive which occasions territorial behaviour it will help us to understand its meaning and to elucidate what is fundamental and what derivative in its various manifestations.

Howard eschewed furnishing a definition, but various formulae have been suggested by Mayr (1935), Lack (1939 a), Tinbergen (1936 b, 1939 a), Davis (1940 a) and others. Lack's definition is as follows: 'An isolated area defended by one individual of a species or by a breeding pair against intruders of the same species, and in which the owner of the territory makes itself conspicuous.' This approximates to Mayr's definition apart from the inclusion of the female. Mayr holds that territory was originally developed only in connexion with mating, but has acquired in certain Passerine species a secondary significance as the food-providing area. In his paper on the snow bunting Tinbergen abandons the attempt to find a formula which will usefully describe all kinds of bird territory and restricts his definition to sexual territory. It is 'an area that is defended by a fighting bird against individuals of the same species and sex shortly before and during the formation of the sexual bond'. He considers that, 'Whenever sexual fighting is confined to a restricted area, this area is territory.' As Tinbergen himself points out that there are exceptions—such as the wren-tit (Erickson, 1938)—to the rule that in defending territory birds of like sex oppose each other, it is not easy to understand why such birds are excluded from his first definition. He thinks it wrong to include any reference to the function of territory in a definition of it as the function varies according to species, but in so far as it has to do with sex, according to his own definition, it is concerned with reproductive function.

Davis, attempting a general definition says, 'Territorialism may be described as the defence of an object (territory) which serves in reproduction.' This, it will be noted, is inclusive enough to cover the case of the male bitterling (*Rhodeus*) which defends the perambulating mussel which is the receptacle for the female's eggs. The definition is designed to include the four main types of freehold—mating and nesting areas, locations useful for the feeding of the young and winter habitats. Exception might be taken to it on the grounds that the term 'territory' should be applicable to areas which birds (or other animals) defend because there they obtain their own food supply, but it is, perhaps, best to exclude such feeding areas by definition. It will be noticed that all these formulae stress the fact that territory is relevant to reproduction. Viewed psychologically territory is concerned with sex rather than with hunger. It is an area essential to the fulfilment of the sexual cycle, and a considerable part of its value to each pair of birds lies in the comparative immunity from disturbance of their sexual functions which it secures. The freehold is usually more the concern of the male than the female, though she often co-operates in defending it and female robins and mockingbirds may hold territories. Commonly the cock demarcates the boundaries and his mate learns them from him, sometimes imperfectly, as observed in willow wrens (Brock, 1910). A female song sparrow (Nice, 1936) or snow bunting (Tinbergen, 1939a) may even nest outside the territory and render it necessary for her mate to conquer the surrounding area, and a cock blackbird has been known to alter the boundaries of his territory to accommodate it to the changed movements of the female (Dewar, 1920b). If the basis of territoriality were the preservation of a feeding area, the difference in the degree of proprietary concern between male and female would be inexplicable. Still more significant is the fact that the defence of territory does not become more vehement when the young hatch. Some birds feed and gather food for the young outside the territory, others never fight with their principal food competitors, yet others defend an area far greater than is necessary to safeguard the food supply. A multitude of other facts show that amongst birds the sex urge and not the appetite of hunger is the emotional force behind this powerful psychological mechanism; for territory is a psychological construction before it is geographical.

Whatever the various functions of territory may be it will clarify its meaning if we consider the psychological roots from which it has sprung. So ardent are birds in the quest for a territory and so vehement are they in its defence that we may well suspect that some powerful emotional drive must be the source of such behaviour. To discern the nature of this drive is to reveal the true character and meaning of territorialism.

It is of biological advantage to the race that the most intimate and important of functions, copulation, should not be subject to interference, yet birds are much inclined to disturb others in the act. On such occasions sooty terns will assault their neighbours and 'coition is completed only after much fighting' (Watson, 1908). In an aviary, when a robin sees another copulating he will drive him off the female's back (Lack, 1940 d). When prairie hens copulate all the males in the neighbourhood seek to try to dismount their fellow cock (Gross, 1928). Several sage grouse will attack a neighbouring bird seen to be treading a hen (Simon, 1940). Pigeons react in a similar way. A gander who has made himself despot amongst the geese will assault other birds in the act of pairing although he may not be at all anxious to court the goose himself. Heinroth (1911) comments, 'To be scandalised by the sexual acts of others is not therefore a peculiarity of the human mind, but a natural feeling often to be observed in the animal world.' Katz (1937) endorses this view. There is evidence of such feelings of scandalisation in man (Michels, 1914) and possibly in apes

(Sokolowsky, 1923), but the word implies a moral element which we are hardly justified in attributing to birds. All that we can say is that such overt sexual behaviour arouses antagonism or revulsion which in the human species is usually tinged with moral disapprobation. But readers of Captain Cook's diaries will hesitate to conclude that in this matter any particular human reaction is innate. Years ago Selous (1913–16) recorded that rooks are mobbed by their fellows during coitus, but Yeates (1934) maintains that 'legitimate' unions do not call forth more than the excited cawing of two or three birds around the pair concerned. Although supported by Deane (1939), this observation needs confirmation. The important point to note is that a male rook copulating within his territory, that is, on the nest, is not assaulted, but outside his territory he is subject to hard buffeting. We have already noticed that blackcock try to interfere with coitus on the lek. Even displaying birds, as in the case of the herring gulls watched by Tinbergen (1939a), excite much fiercer attack than those not displaying. Sexual jealousy has been shown to be strong amongst birds by many writers. Much of a cock song sparrow's time before his mate has begun to incubate is taken up with preventing her from straying into the adjacent domains of other males (Nice, 1937). The intimate connexion between the nesting area, proprietary rights and copulation is shown in birds such as cormorants which build the nest where the nuptial rite has been consummated (Selous, 1905a, 1927) and grebes which copulate on the nest. Memory is strongest for places where emotions have been stirred (Fischel, 1937). This is an important factor in territory formation.

The possession of territory gives confidence to the breeding male and awes potential intruders. It is a simple and efficient device to restrict violent dysgenic fighting while permitting healthy competition. At the same time—apart from exceptional cases such as the arena-frequenting birds—nest and young are safeguarded. In some instances, as, for example, the communal-nesting smooth-billed ani, sex jealousy has been subordinated and the communal territory in which the nest is situated is defended by all the birds (Davis, 1940a), in others such as the laughing gull the pairing-up and nesting areas are separate. For yet other species the territory appears to be useful pre-eminently as a food-producing area. But there is an overwhelming array of evidence showing that territory is basically concerned with sex and the reproductive functions. Nor is it a serious objection to point out that sexual fighting and jealousy may occur apart from territory, for this is only what we should expect of so powerful a drive.

So strong is territoriality that it might have defeated the very end for which it has been evolved, by preventing the females from entering the preserves of the males, had not elaborate ceremonies, such as we have discussed earlier, been developed to assuage the territory-defending ardour of the male and subdue the inhibitions of the female against trespassing. Even a dominant male manakin will not dare intrude beyond a certain point in the territory of a submissive bird. No impulse so 'psychological' and tenuous as that associated with territory could have gained such strength within the bird's disposition and become so completely innate unless it were of supreme importance in furthering racial survival. Only as a device for protecting the most fundamental requirements of the race can we understand the evolution of a complicated innate psychological constitution, which, were it not so strictly safeguarded, would thwart the individual and result in the decline of the race.

Displaying birds are only a degree less defenceless than copulating birds, and both are liable to arouse feelings of jealous enmity amongst other males of their own species. The territory convention permits display to take place without interference.

And in so far as jealousy for the safety of nest, eggs and young is advantageous to the race, these are included within the orbit of territorial feeling.

We have seen that territory tends to restrict fighting within certain limits, psychological and geographical. It prevents birds too frequently 'getting to grips'. The advertising song often associated with it is a device which serves the same end. The song delimits the territory, but the boundary of the territory also protects the singer in so far as an intruder's compunction increases and his courage diminishes as he trespasses inside it. A singing bird makes himself vulnerable in making himself conspicuous, but the possession of a territory confers a certain amount of protection upon him, for fear of trespassing is innate and not the result of punishment received (Rothschild, 1941).

Besides warning off or repelling rivals, or both, advertising song, in certain species at least, tends to attract a mate. We have seen that the love-call of the heron has this effect (Verwey, 1930). Not only do birds in general sing more vigorously when unmated, but it has been noticed that a quondam silent snow bunting will start to sing even when his mate absents herself for a short time. Probably the song has some effect in impressing the male's dominance on the female.

The deterrent effect of song on other males is shown by Lack's (1939a) notes on a robin:

Song advertises ownership of territory, as propounded by Howard, and, like posturing, often replaces fighting. For instance, on May 27th., 1936, an unringed newcomer started to sing in the corner of a territory owned by male 44. Male 44, then in the opposite corner, promptly sang, the newcomer (which could not yet be aware that it was in occupied territory) sang again, the owner replied from nearer, the newcomer again sang, and this procedure was repeated twice more, the owner finally answering from only about fifteen yards away, still hidden in thick bushes. At this the newcomer fled, from an opponent it never saw, and did not appear again.

Singing is thus commonly display-fighting and is a common substitute for actual fighting.

The connexion between song and territory has been commented on by a host of writers from Olina (1622) onwards. Even earlier Shakespeare had noticed how regularly the nightingale sang from the same place, for Juliet tries to persuade Romeo:

> It was the nightingale and not the lark,
> That pierc'd the fearful hollow of thine ear,
> Nightly she sings on yond pomegranate-tree.
>
> *Romeo and Juliet*, III, v.

When male warblers arrive back in their breeding quarters, preceding the females by a week to a fortnight, they sing from a series of prominent points and thus mark out their holdings. Grasshopper warblers' nests may be found by noticing the spot from which the male first sings and then searching within a radius of twenty paces. The close connexion between song and territory is clear when we consider that although female song is extremely rare amongst birds, female robins, and American mockingbirds sing as well as defend territories in autumn. Years ago Lynes (1913) pointed out the correlation between occupation of territory, precocious drumming and the development of the gonads in the snipe.

It is not surprising that many birds use their powers of flight to augment the advertising efficiency of their songs. Visible and audible display are thus allied. One of the commonest types of such combined display is that in which the bird flies up whilst singing. The whitethroat dancing above the brambles, the tree pipit flinging up from the elm tree or the grouse from the heath, the rock pipit trilling over an islet and the skylark storming the heavens are common examples. The bobolink is equally familiar in North America. In Trinidad the glossy grassquit is continually fluttering up a few feet into the air. *Mirafra fasciolata*, a member of the lark family which inhabits the Congo basin, will shoot up twenty times in succession to a height of thirty feet, make a burring sound with its wings, give a tri-

PLATE XIX

SPOONBILL RESENTING GULL'S INTRUSION ON ITS
NESTING TERRITORY

Owing to flooding the black-headed gull was forced to nest at the base of the
spoonbill's nest. The result was continual bickering between the two species.

PLATE XX

page 7

IVORY-BILLED WOODPECKERS

The 'change-over' at the nest.

page 254

NIGHTINGALES

Male singing and displaying to his mate on the nest.

syllabic whistle and volplane to the ground with wings elevated and legs outstretched (Lynes, 1938). The king bird of paradise has been seen soaring thirty feet into the air like a skylark and then dropping as if shot into the trees (Goodfellow, 1915). The greenshank ascends to a height of as much as 2000 feet with rapid flight, then rises and falls in the air performing evolutions which make the aerobatics of the raven seem prosaic, the while uttering his passionate love-song (Gordon, 1936). The courtship flights of the knot have been described by Manniche (1910):

> The pairing notes of the males filled the air everywhere a few hours after their arrival. The male suddenly gets up from the snow-clad ground, and producing the most beautiful flute-like notes, following an oblique line with rapid wing-strokes mounts to an enormous height, often so high that he cannot be followed by the naked eye. Up here in the clear frosty air he flies around in large circles with quivering wings, and his melodious far-sounding notes are heard far and wide over the country.... The song sounds now more distant, now nearer, when three or four males are singing at the same time. Now and then the bird glides slowly downwards on stiff wings, with the tail feathers spread; then again he makes himself invisible in the higher regions of the air, mounting on wings quivering even faster than before. Only now and then the observer—guided by the continuous song—succeeds for a moment in discerning the bird at a certain attitude of flight, when the strong sunlight falls upon his golden coloured breast or light wings. Gradually, as in increasing excitement he executes the convulsive vibration of his wings, his song changes to deeper notes—following quickly after each other—at last to die out while the bird at the same time drops to the earth on stiff wings strongly bent upward.

According to this account the song contains flute-like notes resembling the love-song of the curlew, but is too rich and varied to be compared with the song of any other bird.

Many different kinds of sound are produced mechanically by wings or tail during display-flight. The noisy flapping of agitated lapwings often appears to have an intimidatory rather than a courtship function and is called forth by intruders of all kinds. Other performances are more definitely in the nature of advertisement; the airy 'drumming' of the snipe and the black-

throated honey-guide (Wood, 1940), the humming glide of the
southern dunlin, the roaring descent of the black-tailed godwit,
and the boom of the American nighthawk.[1] In Guatemala
Salvin noticed that the male black penelope when flying downward with outstretched wings gave forth a kind of crashing,
rushing noise like the sound of a falling tree (Ogilvie-Grant,
1896).

Often, but not always, there is a close correlation between
display-flight and territory. The lark has the spacious and
illimitable skies at his disposal and yet he is tethered by an invisible bond to a patch of ground. He spirals up and at the top
of his ascent wanders in air, but constantly adjusts his flight to
the boundary below. Howard (1929) says:

> Watch then at the start, sometime early in February. For some days we
> may not see the birds, may not see any life, nor hear any song. But a morning
> comes when that which has long been a silent wilderness turns again into
> song. Mark a male as he ascends, mark the limits of his movements in air,
> learn his limits on the ground, and whether spiralling or wandering you will
> observe that he conforms to his territory below, and that even when driven
> by the wind across his boundary he checks himself with palpable effort to
> beat back across it. So that he has two distinct ways of behaving with
> reference to his territory—air movement and ground movement. From the
> beginning air movement is definite and is determined by the ground plan.

With some species, however, such as the black-tailed godwit,
in which the flight-display is more conspicuous than the song,
the bird circles with slowly flapping wings and appropriate call
outside his territory, but always returns to it (Howard, 1920;
Huxley and Montague, 1926). Generally speaking the higher
the development of song the more definitely associated with
territory it seems to be. Far-carrying, elaborately musical, continuous song is usually strictly territorial. Exceptions occur,
however, amongst the Wading birds.

---

[1] The male nightjar claps his wings as a call-note, the female in times of
sexual excitement (Lack, 1932), and the short-eared owl claps as a threat as
well as in nuptial flight.

High-flying vocal or instrumental displays are most characteristic of birds of the open country—moorland, tundra, steppe or pampas. The function of song in delimiting an exact territory may become secondary. The prime consideration is to attract the attention of the immigrant female. But it is significant that one of the few display-flights which has been carefully investigated, that of the red-necked phalarope, has been found to be not only conspicuous, but attractive. 'The ceremonial flight expresses desire for the sex-partner, and has a stimulating (attracting) influence on the other sex, and therefore acts as a means to bring the sex-partners together' (Tinbergen, 1935).

Amongst birds of the open spaces any lack of continuity or penetration in the song is remedied by conspicuous, wide-ranging, sometimes noisy, flight. Most waders, as, for example, curlews, snipe, sandpipers, stints, plover, redshanks and greenshanks execute such display-flights. Some nightjars, such as the American nighthawk, and even owls, when they adapt themselves to open country as the short-eared has done, acquire a song-flight (Dubois, 1924). Deficiency in song tends to be counterbalanced by conspicuousness in flight.

In some species in which song is not highly developed the function of the display-flight seems to be to stimulate the female rather than to attract her to the territory. In spring and summer male oystercatchers may be seen flying around with slow wing-flaps, butterfly-fashion, followed by the female.[1] The magpie indulges in a beautiful display-flight which I have not seen described. Early in the year when I have noticed three birds together the males utter a rapid clinking cry—a higher-pitched version of their usual chuckle, which it sometimes immediately

---

[1] Howard (1929) has pointed out that two forms of sexual flight amongst birds may be distinguished: the butterfly and the moth types. In the first the fully expanded wings are flapped slowly; in the latter the partially opened wings are rapidly vibrated. The one is characteristic of godwit, oystercatcher and guillemot, the other of the reed bunting.

follows. After perching on some lofty point for a time uttering a medley of these calls the male will set out on a high direct flight, the female accompanying him some distance away. Every two or three seconds as he flies he gives a double or treble rapid beat of his wings, a curious intermittent flutter, comparable in effect to the leg-flutter of a certain style of crawl swimming. The performance takes place approximately over what will become the breeding territory, but not at an habitual display-ground.

Many are the varieties of display-flight, from noisy vocal or instrumental performances to silent evolutions. Amongst birds of prey the wing display is often impressive. The kestrel swoops again and again with terrific speed, checking the force of his descent when he is close to the branch on which his lady-love is perched, or hovers over her with rapidly fluttered wings and excited squeals. The sparrowhawk flaps high aloft with curious, deliberate, jay-like flight, then drops like a stone, and repeats this again and again. Such a swooping display-flight is characteristic of many birds of prey. Kites circle in pairs, now one, now the other dropping abruptly to zigzag over the surface of a lake (Rudolph, 1898). Sick (1936) observed a most remarkable dual performance by two black kites. The pair clasped each other's claws and remained poised in mid-air with beating wings for some seconds in an upright position. As they began to fall they started to revolve round their interlocked claws as axis, at first slowly, then faster, but with amazing regularity and never so rapidly as to suggest loss of control. On four occasions the birds nearly touched the ground before disengaging themselves. Walt Whitman recorded a similar display:

Skyward in air a sudden muffled sound, the dalliance of the eagles,
The rushing amorous contact high in space together,
The clinching interlocking claws, a living, fierce, gyrating wheel,

Four beating wings, two beaks, a swirling mass tight grappling,
In tumbling turning clustering loops, straight downward
    falling,
Till o'er the river pois'd, the twain yet one, a moment's lull,
A motionless still balance in the air, then parting, talons loosing,
Upward again on slow-firm pinions slanting, their separate
    diverse flight,
She hers, he his, pursuing.

The Dalliance of the Eagles.

Aerial sexual pursuit is, of course, a common feature of bird life
and in some species marks a significant phase in the reproductive
cycle (Howard, 1935). Probably it represents a primitive,
stimulatory form of courtship and, no doubt, it is the source of
various types of ceremonial flight. Like so many other forms of
display it tends to become social. When the most intolerant of
all activities, the defence of territory, can become communal,
it is not surprising that the same thing can happen in the case of
display-flights. In many species the sight of three birds in close
pursuit of one another is a common sight. Display-flights of
dunlin, golden, grey and ringed plover may become social
affairs. Four or five sociable plover will chase a female. Neighbouring breeding nightjars (Lack, 1932) and bitterns (Turner,
1924) engage in play-flights without showing any territorial
aggressiveness. A distinction must be made, however, between
social aerial displays such as the aerobatics of humming-birds
which have a competitive basis and flights which are largely
sociable and playful like those of the crested screamer.

In courtship Anna's humming-bird descends with great speed,
then turns sharply upwards with a chirp and sings for a few
seconds directly above the female (Woods, 1940). Allen's
hummer flies slowly to and fro like a giant pendulum (Woods,
1927; Bent, 1940)—a love-flight similar to that of the finch
which Azara aptly called *Oscilador* (Hudson, 1892). The broad-

tailed humming-bird swoops back and forth before the female in a great open U, the arms of which may be fifty or sixty feet high (Skutch, 1940).

Humming-bird courtship displays may be grouped into two extreme types, the static and the dynamic. The species just mentioned exemplify the dynamic type. In the other group are the species in which the male establishes a headquarters which he frequents during the breeding season. Alone on a favourite perch he reiterates an unmelodious call to attract the female's attention. But often two or more birds assemble and 'sing' together (Skutch, 1940). It is undesirable to refer to these assemblies as leks, for as yet evidence is lacking to show that the behaviour of the birds is comparable with that of blackcock.

In dynamic display more than one humming-bird may take part. This is the case with the great Jacobin according to Belt (1874). First one and then another of two males perform in turn before the female as she perches on a branch—flying aloft, then descending slowly with the white tail expanded so widely that it covers more space than all the rest of the bird. Aerobatics in which several birds take part are also recorded of the genus *Phaethornis*. The racket-tailed humming-bird executes elaborate aerial antics:

> Two birds would first of all hover in the air opposite to one another with their bodies in a vertical position, then with expanded tail they flew first to one side and then the other uttering a clicking note. Sometimes several males took part in this performance. At other times one bird would hang below a slender branch as another bird went through the mazy dance, when suddenly the position of the two performers would be exactly reversed, one taking the position and performing the actions of the other. Meantime the wonderful racket-shaped tail is moved in a variety of ways, the spatulate ends sometimes being bent forwards almost to the crown of the head. (Taczanowski and Stolzmann, 1881.)

This bird has enlarged quills like those of the manakin, enabling it to make a clicking noise when in flight.

I have described the spectacular social flights of various birds, such as the starling and dunlin, elsewhere (1940). Pleasure in corporate activity appears to be the dominant motive. With species which, as pairs, carry out aerial social displays it is possible that the birds are stimulated sexually in thus associating in a joyous frolic together. One example of this kind of display will suffice. The scissor-tails of La Plata always live in pairs, but at sunset several couples assemble, calling excitedly to each other. They then mount like rockets to a great height, wheel about for a few moments and dash downwards in a wild zigzag, opening and shutting the long tail feathers like a pair of shears and producing loud, whirring sounds. Having concluded their aerial dance the separate couples alight on the tree-tops and male and female join in a duet of rapidly repeated castanet-like notes (Hudson, 1892).

Whatever the biological purposes which may be served by such aerial frolics as have been described, some of them seem to incorporate the element of play. How else, indeed, can one describe such other antics as eider ducks shooting rapids again and again as described by Roberts (1934) or the 'cat and mouse' game with their prey which grebes (Hudson, 1920), cormorants and hawks (Finn, 1919) have been seen to carry on. The playful tricks recounted of ravens (Pitt, 1927) and other birds (Groos, 1898; Finn, 1919) are legion, and no naturalist can dismiss them as merely on the anecdotal plane. Nice (1937, 1938) has shown that although the main functions of the song sparrow's song are to defy other males and call the female, yet he sings while the female is incubating partly because of his super-abundant energy. Surplus vitality may manifest itself in many other ways and it is reasonable to believe that communal aerobatics, like water-dances such as those of black guillemots, partake of the nature of sport, and that not all bird activity is of a utilitarian character.

CHAPTER XVIII

THE SIGNIFICANCE OF DISPLAY

Correlation between adornments and display—Distinctiveness in song—
The theory of releasers—Confirmation in spider display—Analogy of
traffic signals—Stimulation of sexual or aggressive behaviour by morphological features—Explanation of apparent necrophilia—Mating-types—
Warning and flying-in-pursuit releasers—The releasers of importunate
nestlings—Imprinting—Experimental work—Influence of physiological
condition—The bird's world as a dynamic system.

THE element which more than any other is characteristic
of sexual display amongst birds is strangeness. Attitudes
and movements tend to be odd, exaggerated or unwonted. Peculiar adornments are thrust into prominence:
crests, wattles, ruffs, collars, tippets, trains, spurs, excrescences
on wings and bills, lurid mouths, tails of weird or exquisite
form, bladders, highly coloured patches of bare skin, elongated
plumes, brightly hued feet and legs. Such features as the blue
horns of the crimson horned pheasant, and Bulwer's pheasant
(Heinroth, 1938 a), the strange erectile wattles arising from
the base of the Costa Rican bell-bird's beak, the scarlet pouch
of the frigate bird and the yet more immense pouch of the
Australian bustard are exhibited in love-making. The display is
nearly always beautiful; it is always striking.

There is a conspicuous correlation between the movements
of the displaying bird and his special charms of form or coloration. The blue-footed booby brings his feet into prominence,
the Bahama pintail puts his head under water and jerks up his
tail, the Amherst pheasant erects his neck-ruff on the side next
the female. The goldfinch strikes attitudes in his wooing different
from those of any other British finch, but perfectly adapted to
make the most of his adornments. Expanding his exquisitely

decorated wings he turns quickly from side to side before the hen.

> Sometimes goldfinches one by one will drop
> From low-hung branches: little space they stop
> But sip and twitter, and their feathers sleek;
> Then off at once, as in a wanton freak:
> Or perhaps, to show their black and golden wings,
> Pausing upon their yellow flutterings.
>
> Keats, 'I stood tip-toe upon a little hill.'

The closely related siskin does not posture thus, for he has not the beautiful wings of the goldfinch to display.

The diversity of bird voices is no less than the variety of form and colour. There is hardly an instrument of the orchestra whose tones are not simulated by some love-sick bird. Storks play castanets; pheasants, ruffed grouse and woodpeckers beat the war and wedding drum. There are innumerable performers on flute, piccolo and oboe. Cranes trumpet, geese sound horns, the nightingale plucks his golden harp-string and the oropendola his zither, the bell-bird clangs his eerie bell in the jungle and the little bittern pile-drives in the swamp.

It should be noticed that song and call differ in closely related species, such as the chiffchaff and willow warbler, much more than posturing. The importance of this is evident when we consider that posturing is a device to elicit response when male and female are relatively close together, but song must be effective at a distance to be effective at all. Writing on this subject Howard (1929) says:

Despite all the differences we can detect—and they are great—we recognise the owner of the voice without difficulty. For throughout there runs a specific refrain which is sufficiently distinct to elicit reaction in the female belonging to the species. Elaboration, tone, pitch, modulation—these count only in so far as they contribute towards distinctiveness.

This remains true even of song sparrows in which there is great individual difference in song. Many birds, however, have specific pairing notes which are for use at close quarters and are therefore more subdued than the notes of the song.

In the foregoing pages we have had to eschew anything but the briefest reference to the multifarious colours, shapes and sounds which are characteristic of displaying birds, and here we can only comment upon them as contributing to the whole pattern of display-behaviour. Colour, sound, form and function are integrated so intimately that to scrutinise one aspect of the performance, as we sometimes have to do, is apt to compromise our conclusions. With this warning in mind let us enquire how various movements of the male bird in courtship and the exhibition of specific adornments or utterance of particular calls influence the female, and further pairing-up and procreation.

A good deal of the discussion about the meaning of bird adornment and display has been vitiated by the unconscious assumption that the bird perceiving and affected by the display is, like the ideal ornithologist, an open-minded, objective observer. A moment's reflexion shows that this cannot be so. The relationship is much more that of wireless transmitter and receiving set. The transmitter operates in vain unless there is an apparatus adapted to its wave length prepared to receive its signals. So unless there is in the avian spectator's organism some adaptation towards receptivity, some complementary psychological mechanism, the other's adornment is meaningless and the display will be in vain.

The theory of the correlation of adornment and response set forth by Lorenz (1935, 1937, 1939) has been generally welcomed because it goes farther than any other hypothesis towards explaining the significance of the wide range of avian decoration and ceremonial activity. It involves the conception of 're-leasers'—postures, movements, adornments and sounds—which have the power of arousing or 'switching on' specific reactions in the 'associate' bird. Lorenz has shown that in the appropriate conditions 'releasers' call forth the response with an inevitability as potent as the 'open sesame' of the fairy tales. This being so it is of supreme importance that they should be, so far as possible, unmistakable and not to be confused with anything else which may appear in the environment or any posture with different significance which might be adopted by an associated bird. Were it otherwise, the bird would live an Alice in Wonderland existence, continually arousing inappropriate reactions, thwarting its own and other birds' urge to pair and breed, and constantly acting on false clues to the jeopardy of its life.

The importance of easily recognisable clues is revealed by a study of spider behaviour. The Attidae and Lycosidae are hunting spiders endowed with relatively good vision which secure their prey by springing on it from a distance, and the sight of a small, moving object is sufficient to set the preying reactions in motion. Thus it seems that the struggles which take place between males are due to mistaken identity rather than rivalry or 'jealousy'. Errors of recognition also occur between male and female. Bristowe (1929) writes:

Spiders can see moving objects at a distance outside their range of distinct vision, and as the male approaches the female, who is larger than himself, he stands in considerable danger of being killed or injured. Now I have found in the course of my observations that once a male is recognised by the female he is relatively safe, even though she will not submit to his advances at once... so if a male were to acquire any distinctive variation which tended to make him more easily recognisable, he would stand a better chance of survival than his fellows... the more distinctive the leg-raising and the display of the decorations, the less the danger of the male.

If a maladroit lover does not declare himself as such by a prompt and appropriate display or dance the female may mistake him for insect prey, leap upon him and bury her fangs in his body. To avoid this he must act as no insect ever would, conforming to the traditional pattern of display, exhibiting his peculiar markings and generally doing what is necessary to achieve recognition and release the sexual reactions of the female. Thus the male starts his antics, which often amount to a dance, on the evidence of scent alone. His behaviour inhibits aggression and stimulates sexual response.[1]

[1] An illuminating study might be made comparing the displays of spiders with those of birds as the following notes will show. In an earlier chapter I showed that the releasers of male spiders played an important part in saving them from being devoured by the females they approached. A spider's adornments have a vital function as identificatory clues, but they also operate in stimulating the sexual responses of the female in a way very similar to the releasers of birds. Having achieved immunity from the damsel's fangs the male spider must arouse very different behaviour from that which any other small living creature usually excites. What happens is best described in the words of Bristowe (1929):

'Once a male has started his displays he is relatively safe, but before the female will accept him he must stimulate her. She possesses a very strong "preying instinct" and her sexual instinct must be aroused to a point where it dominates her desire to kill before she will submit her body to him. The amount of stimulation required varies considerably in the same individual from time to time and in different species, but in due course she submits and allows him to approach her—in some cases·reciprocating his movements. The female watches the movements of the male, and there is no doubt his display has a stimulating effect on her. It is probable that before copulation *can* take place the female must be stimulated, and until the necessary amount of stimulation has been effected, copulation is physically impossible.... The display by the male is essential to the species, and therefore Natural Selection is certainly involved. The probable evolution of the display and the male epigamic decorations, coupled with the fact that the importunity of different males appears to vary appreciably would appear to make it likely that the female's reactions to different males is not exactly the same, and as her sexual desires begin to wane, it will not be every male which succeeds in exciting her sufficiently to secure her submission.'

It is, perhaps, of interest to quote the details of an actual performance, as it illustrates the almost compulsive character of the stimulation and provides an interesting comparison with the display of the shag as observed by Selous and cited earlier:

'When placed with a female (*Euophrys frontalis*) the male raises his hand-some front legs with a jerk, accompanied by a jerk of the abdomen, to an almost vertical position above his head, and then slowly lowers them until they almost or just reach the ground, when they are snapped up to the

The feeding reaction of a mackerel is released by an object which has two characteristics—a silvery, glistening surface and motion through the water. This fish is effectively equipped for hunting its prey by the very simplicity of the releaser required, for the delay consequent upon close scrutiny of the shape and markings of potential prey would entail many a lost meal. Man, however, is able to exploit the situation and has discovered that a strip of skin from a dead mackerel on a hook drawn through the water will deceive the living fish. Birds may even deceive other birds unintentionally. The smooth-billed ani has a distinctive alarm note which releases flying-up behaviour, but the mockingbird has incorporated it in its song and unintentionally causes the anis to fly up when no danger threatens (Davis, 1940 a).

Indian house crows show just as much sympathetic excitement and use the same cries and gestures when a living or dead bird of dark plumage is handled as when one of their own kind is captured. Even a drongo shrike, which preys upon crows, will arouse demonstrations of sympathy (Finn, 1919). Other members of the Corvidae have equally simple releasers. Lorenz (1938) was one day carrying a wet, black bathing costume when his tame, free-living jackdaws suddenly mobbed him. He investigated the reason for this surprising attack and his experi-

vertical position once more and then lowered. This procedure is repeated over and over again, though sometimes varied by two snaps in rapid succession, and the male usually advances a step between each upward snap of the legs. At the same time the yellow palpi, which are bent inwards tip to tip are vibrated up and down in front of his black face. These mesmeric movements are repeated over and over again, and sometimes the female roughly reciprocates by movements of her palpi and the raising and lowering of her front legs. She appears to be very much influenced by his courtship, and one female which was almost ready to lay her eggs and was unwilling to mate with him seemed loath to tear herself away. Several times she turned away, walked a short distance, and then turned round to watch him again. Only his close approach made her move away, and on one or two occasions she imitated his movements.'

ments showed that anything black, glistening and dangling carried by any living thing elicited the reaction. He could handle young naked jackdaws with impunity, but was assaulted by the adults the first day that black quills were visible on the nestlings (1938). Even jackdaws carrying nesting material approximating to this description are mobbed by their fellows. The birds are so ready to act on a basis of communal responsibility that they attack any creature which seizes one of their number, but although they readily recognise each other individually in the flock it is not these recognitional factors which come into play in emergencies. The social defence reaction is aroused in a much simpler way. It is by the suddenness and simultaneity of their onslaught that they may best rescue a fellow in distress. Nature eliminates the non-essential as a hindrance to quick and effective action and a handicap to the race. As a rule 'blitz' methods are most successful in attack or defence.

Given a certain restricted but definite set of combined stimuli a creature reacts in the characteristic way. We find an analogy in our system of traffic control; a red light brings every vehicle to a stand-still, a green light sets them in motion. Of all the sights and sounds at a busy cross-roads only these have power to bring about such responses. There may be many objects of red and green in the shop windows, many passers-by whose dress displays one of these colours; there may even be lights of red and green in the streets, but the motorist is not misled by them because he knows that a coloured light as a component of a particular pattern of objects at a street corner means a certain thing and calls for a particular response on his part. If some mad practical joker were to erect a set of coloured lights similar to traffic signals outside his shop there would be utter confusion until the police intervened to put matters right. Motorists would be tricked like the spider, mackerel or jackdaw. The traffic-conscious mind acts on the necessarily unquestioned assumption that a set of coloured lights in a certain relationship cannot be meaningless or accidental. It is conditioned to this state of affairs, for this is conducive to survival and pacific social relations. The motorist can no more take account of possible lunatics erecting unauthorised signs than the jackdaw can be on his guard against deceptive black bathing suits. The effectiveness of the traffic signals depends on their singularity. That particular kind of combination is highly unlikely to occur elsewhere than at the appropriate place. It is in this way that certain combinations of objects, movements or colours elicit their response from the bird, but motorists differ from birds in that they become conditioned by experience to traffic-signals while birds are endowed with an extensive innate equipment of releasers and releaser-correlates.

When a stuffed female yellowthroat was placed in the territory of a wild male he copulated with it, but when a black paper mask—the male's most conspicuous character—was added he began to copulate but suddenly broke off. He came back to attack the specimen, but when the mask was removed copulated again (Noble and Vogt, 1935). The experiment which Noble (1936) conducted on the flicker, which has already been referred to, revealed a similar state of affairs in regard to that bird's 'moustache'. The male mounted the female when she adopted the invitatory posture; then, seeing the artificial moustache with which she had been provided attacked her as if she were a rival male. Posture and moustache are both sex-recognitional features in this species. Opposing males spread their tails widely, exhibiting the bright yellow under-surface. The coloured tail thus obviously exercises an intimidatory function. It is an interesting point that the difference in hue of these feathers does not prevent the inter-breeding of the red-shafted and yellow-shafted flickers. Similarly the goldfinch and the grey-headed goldfinch of the Himalayas inter-breed where their ranges over-lap. Experiment with these species would probably show that the colour of the head is not sex-recognitional, just as the tail-feathers have not that function in the flickers. The wing-display, as we have already seen, is most important in goldfinch courtship. Head colouring is often functional in display. Jenner Weir told Darwin (1871) of attacks made on a crossbill and a goldfinch by a robin and on a reed bunting by a bullfinch when the bunting acquired the black head-dress of full breeding plumage. Similarly an indigo bunting scalped a painted bunting, which also has a blue head. A mated superb blue Australian wren so persecuted a cock yellow-winged blue sugar-bird that it had to be removed from the aviary (Finn, 1919).

It should be noted that where species are much alike in plumage, but have different notes, as, for instance, the willow warbler and the chiffchaff, interbreeding in natural conditions seldom occurs. It is much commoner where the distinction is in feather coloration. The importance of the call in recognition hardly needs to be stressed. The gentoo penguin recognises its mate by the sound of its crowing. The female flicker has a special invitatory note—so has the female song sparrow. The female great spotted woodpecker during her invitation delivers three series of three rapid strokes on a branch or tree-trunk.

Sexual releasers employed by the hen bird constitute more striking examples than those used by the cock in that often primary sexual reaction is almost instantaneously visible. The immediate stimulus to coition, according to Lorenz, nearly always comes from the female, and the distinguishing mark is often squatting and remaining motionless, or, in some cases, fluttering the wings. This is the explanation of the fact that a great many species which have been subjected to experiment, varying from the nuthatch to terns, will endeavour to copulate with a stuffed specimen. Allen (1934) found that a male ruffed grouse would mount a dead grouse or a grouse skin so long as the

specimen was more or less flattened. Audubon (1840–44) wrote
of the eastern turkey that when two fight and one kills the other,
'the moment he is dead, the conqueror treads him under foot,
but, what is strange, not with hatred, but with all the motions
he employs in caressing the female'. Such behaviour by
partridges was commented upon by Aristotle and Pliny. Penguins
(Bagshawe, 1938), redstarts and robins (Heinroth, 1924–33) try
to copulate with dead members of their own species. Similar
responses have been noted in fish. The female stickleback when
ready to lay her eggs swims towards the male with her head
upwards and body in a semi-vertical attitude—a most unusual
position. Experiments have shown that dead males or females,
and even dead fishes of other species floating in this way provoke
the male stickleback's response (Tinbergen, 1939 b). This kind of
behaviour is not necrophilia. A fish can hardly do anything more
singular than swim vertically nor a creature so volatile as a bird
act more unwontedly than to squat motionless. These signals
are as definite and unmistakable as possible and conform to the
requirements of releasers as defined by Lorenz, combining
simplicity with singularity. The infrequency of false responses
shows how effective they are, and when mistakes occur they are
useful in revealing the true nature of the mechanism involved.
Releasers may be in series, as when the male snow bunting reacts
with threat display to an approaching bird and if it fails to reveal
itself as a male by reciprocating with aggressive behaviour the
invitatory display is then given.

Repeating and amplifying an earlier reference in these pages
let us turn to Lorenz's classification of mating types amongst
birds. Using as criterion the male's mode of sex recognition he
distinguishes the 'lizard', 'labyrinth-fish' and 'cichlid-fish'
types. The names are clumsy but useful, as lizards and some fish
show clearly defined forms of display. In the lizard type the
male displays his bright coloration and attempts coition if the
other does not show a similar display. In the labyrinth-fish type

the display of the male develops into attack unless the response is characteristic female behaviour. The female shows stimulative behaviour, but the threatening display of both male and female in non-sexual conflicts is the same as the male sexual display. In non-sexual struggles the male's submissive behaviour is the same as the female's sexual display. Thus certain types of apparently homosexual behaviour are accounted for. The cichlid-fish type includes those species of birds in which male and female display is alike or very similar. Both sexes display releasers, though their display is not necessarily identical. Thus, in the lizard type the male alone shows mating releasers, in the labyrinth-fish type the female has releasers and individuals who do not show female releasers are attacked. In the cichlid-fish group both male and female have releasers. Tinbergen (1939 a) has shown that this triple grouping requires considerable modification. In attempting to classify mating types on the basis of the method of sex recognition Lorenz has had to use other criteria, such as the dominance relationship. We have amongst birds a series of intermediate mating forms between the two extremes, the lizard and the labyrinth-fish types, approximating sometimes to the one, sometimes to the other. Both sexes have releasers which operate at pair-formation, but according as the releasers in one or the other sex are vaguely defined the display approaches the lizard or the labyrinth-fish type. Lack (1940 e) thinks that pair-formation, especially amongst non-migratory and flock-pairing birds, is a much more complex matter than Lorenz's classification would suggest.

It must not be thought that releasers operate only between male and female. A bird is furnished with correlates to significant objects in its environment of various kinds, animate and inanimate. Lorenz (1939) says that the whole sociology of higher animals is built on releasers and innate patterns. Social releasers have an important function amongst birds which associate in flocks. The grey lag goose has a beautiful grey and black pattern on the wings so that when flying this design is displayed to the succeeding bird in the gaggle. We can interpret the fact already mentioned—that large flocks of geese fly up oftener than small parties—as due to the stimulatory effect of these markings. Their function is to promote a flying-in-pursuit reaction and they may be compared with the white scuts of does which serve to arouse a similar response in their fawns. The value of these releasers is more than their value as warning signals. They switch on the response in their fellows so that the reaction to danger is very rapid. Flying-in-pursuit emblems such as the white bar at the base of the tail in Wilson's petrel may also serve to provoke sexual flight (Roberts, 1940*b*).

Another type of releaser concerns parent birds and their young. Examples of these have already been given in our discussion of the mouth coloration of nestlings. The movements of their heads also may have a releasive function. Begging-releasers may be movements up and down as in parrots, gulls and spoonbills, sideways as in finches, rotary as in European orioles (*Oriolus*) or rapid trembling as in warblers of the genus *Sylvia* (Lorenz, 1937). Lorenz shows that the cheeping of a duckling in distress is the signal which puts the defence reaction of the mallard duck into action, but it is the pattern of the duckling's down which is her recognition signal. Thus she will rush to rescue a Muscovy duckling which cheeps like a young mallard, and having done so will attack and even kill it.

Thus releasers function in relation to the bird which is the appropriate 'opposite number'. It is in this connexion that Lorenz uses von Uexküll's term *Kumpan*, or, writing in English, 'companion'. The German word carries the sense of an associate in a particular activity, so the English is less precise. 'Partner' would perhaps be a more suitable term. A bird may be to another in the relationship of parent-partner, child-partner, sex-partner, social-partner, brother- or sister-partner. It possesses a variety of 'keys' consisting of colours, movements and sounds in various relationships which 'unlock' responses in the appropriate partner. It should be mentioned that some birds, such as the grey lag goose, are susceptible at a very early age to what Lorenz calls 'Prägung' or 'imprinting'. If a gosling on hatching sees and follows another creature than its parent it becomes 'fixed' so that it afterwards regards this creature as its natural partner. Thus the gosling may come to regard a man as if he were a goose! (Heinroth, 1911.) Hence the instances recorded from the time of the philosopher Lacydes onwards of geese which followed their masters as faithfully as dogs (Pliny, x, 27). But Lack (1941 b) has pointed out that in the case of the grey lag goose and other birds the effect of 'imprinting' is not necessarily permanent and irreversible. A very similar phenomenon to 'imprinting' has been noticed in insects (Thorpe, 1938, 1939). Whether human fixations come into the same category must be left to the psychologists to determine.

The releaser or constellation of stimuli needed to set behaviour in motion is like a combination lock which requires a definite series of manipulations to open it—a series which it is almost impossible to find by chance. If the key or key set of movements be compared with the stimuli, the lock itself represents the perceptory correlate—that is the potential receptivity of the bird to those stimuli. This is what Lorenz calls the 'innate perceptory pattern'. He says: 'The relation of the particular form of the lock to the key that fits it, or of any innate perceptory pattern to the set of stimuli to which it responds, is ever a compromise between greatest possible simplicity and greatest possible improbability. The improbability of the innate perceptory pattern is to guard the instinctive reaction from being released by chance through other than the biologically "right" influences' (1937).

Movements or adornments used in bird ceremonial operate effectively because of the innate perceptory patterns possessed by the associate for whom they are intended. The display must follow the specifically appropriate pattern or it will be ineffective. The object of the display must be susceptible to that particular ritual or it will all be in vain.

> By no endeavour
> Can magnet ever
> Attract a silver churn!
>
> W. S. Gilbert—*Patience*.

Hence the evolution of display, so true to type in each species, so delightful in its diversity as we pass from species to species—though often a common pattern may be traced underlying the posturings. This ritual is a code so artificial that its purpose can hardly ever in the ordinary course of events be mistaken by those for whom it is intended; an ancient, esoteric, dramatic language which says by gesture that for which we human beings have to employ many words.[1] It may be compared with the marine code of signalling with flags. The shapes, colours and designs are endowed with meaning according to the combinations of the differently coloured scraps of bunting. To the landlubber they are no more than a bright decoration to the ship. At best he will be aware that they have a meaning without being able to detect what that meaning is. The seaman signaller sees them, not as bright adornments of the vessel—though he may, incidentally, have an eye to their beauty—but as meaningful symbols stimulating him in some cases to immediate and drastic action. But the mediation of the meaning of the symbols intellectually makes a distinction between this kind of signalling and the releasers employed by birds.

---

[1] Cf. the dance by means of which bees, returning to the hive, indicate their discovery of a source of nectar (von Frisch, 1937; Françon, 1939).

This illustration is, however, inadequate to represent what takes place when posturings are repeated for weeks or even months on end. Changing slightly the analogy drawn earlier we may compare continued male display to the female with a film radiated by television to a single receiving set. The transmission is ineffective unless sent out on the wave-length to which the receiving apparatus is adjusted—what corresponds in the female bird to a combination of innate perceptory pattern and physiological maturity. The stereotyping of the display may be compared to this stabilised wave-length. But when the pictures are transmitted and received on the correct wave-length there is an effect, more or less immediate, more or less complete. As the film moves to a climax so may a bird be stimulated by successive displays and by the whole dramatic effect of the male's courtship to rise emotionally in response. How fatuous the tail-end of a film seems when we enter the cinema just at the conclusion of the showing of the big picture! How full of meaning and emotional power if we sit through the whole of it!

Tinbergen (1939 a), in criticising Lorenz's views, urges that it is inaccurate to speak of the releasing mechanism or 'Schema' being linked with a sex partner or social companion. One can only speak of the mechanism concerned in a certain reaction. Moreover, the first reaction of some male spiders, squids, fish and birds in the appropriate breeding condition to an approaching individual of the same species is, as we have seen in the case of birds, the same whether it be male or female; but if both be presented at the same time, the male will show slightly different responses to them, indicating that there is differentiation involved in the releasing mechanism. Tinbergen prefers the general term 'signal' to Lorenz's 'releaser' as he thinks the concept 'signal' should be divided into 'releaser' and 'director'. The danger call of the lapwing may have a primarily releasive function, whereas the white tail markings may have a primarily directive function.

Lorenz (1939) has himself accepted these criticisms and agrees that Tinbergen has established the existence of separate instinct-releasing and directing stimuli. He also deplores the practice of referring to the 'innate perceptory pattern of the sex partner' and so forth. He says: 'One cannot properly say that the sum of the receptoral correlates connected with a given object yields a "whole-pattern" (*Gesamtschema*) of the object. Experimentally, every reaction, dependent on its own releasing pattern, is completely independent of all others; and it is releasable by the stimuli specific to it, independently of all other stimuli gathered together in the same object.'

Lorenz and Tinbergen, experimenting together, were able to determine the exact characteristics of a bird of prey which occasion alarm in geese and ducks quite apart from experience or the example of adults. They discovered that the innate releasing patterns of ducks had reference to falcons and that birds with long beating wings such as swifts could arouse the seek-safety mechanisms. For geese, however, the essential feature of a flying alarm-object was slow movement in relation to size. Thus pigeons sailing against the wind caused apprehension so long as they did not beat their wings, and even a floating feather received meticulous scrutiny. They inferred the correlate of the inborn perceptory pattern to be the sea eagle (*Seeadler*).

Birds will react to models which contain the necessary releasing features even although the models bear hardly any resemblance to the objects normally showing the releasers, as when a glistening black bathing costume arouses the social defence reaction of jackdaws or a robin attacks a few red feathers. Clearly responses having reference to inborn patterns which work on the lines of unconditioned reflexes operate in a very different manner from those caused by conditioned releasing situations.

Although Lorenz admits that innate perceptory patterns, such as the gaping pattern of young thrushes which can be transposed into different absolute sizes (Tinbergen and Kuenen, 1939), have some of the characteristics of what is known in psychology as a 'Gestalt', he believes that the effect of a releasing combination is additive and shows that this can be proved experimentally. Characteristics may be subtracted from a given releasing stimulus-combination without preventing it from operating, but its fully effective working depends on its sending out a summation of the relevant stimuli. Movement-characteristics are much the most important in releasing mechanisms.

These theories have the great merit that they provide a basis for experimental work. They explain facts which hitherto have not been susceptible to interpretation. How is it that an octopus does not recognise a crab dangled on a string (Bierens de Haan, 1926 b) and a penguin does not perceive a fish out of water to be an article of food (Roberts, 1940a)? The answer is simple: neither of these objects displays the relevant signals. Bierens de Haan (1929) remarks that 'animals in general react to a whole situation, and often show a curious incapacity out of such a complex to isolate fragments that must be of great importance to them and are unable to recognise these fragments in other complexes', and Russell (1938) has emphasised that animals respond to a functional environment, but Lorenz has carried analysis to the point of being able to say *why* they do not isolate fragments of a complex—as in the case of the jackdaws and the bathing-costume—and has shown *how* the environment is functional.

The successful operation of releasing stimuli depends on the physiological condition of the recipient. The internal state of robins varies in intensity with different birds and in the same bird from day to day. Sometimes it is so strong that intolerance is shown to other species; occasionally an intruding robin is not attacked (Lack, 1939a). The female red-necked phalarope welcomes a newcomer by flying towards it, whether it is male or female, performing a 'ceremonial flight' and uttering a love-call. Mistaken recognition sometimes occurs and a Lapland bunting is greeted as if it were a phalarope, but Tinbergen (1935) noticed that these reactions to strange birds were observed only during the last few days of the unmated period and was thus led to infer that the power of discrimination is affected by strong sexual desire. Possibly this explains another instance of mistaken identity recorded by Selous (1901-2). He saw a cock lapwing make a scrape before three stock doves, and then, suddenly realising his error, stop 'with a little start'. Experiments by Lorenz (1939) throw light on this kind of situation. A mallard will react to a model of a flying bird (in itself neutral) either as if it were a threatening bird of prey or a male of its own species according as to whether the threshold of flight behaviour or courtship behaviour is lowest at the moment. This principle is part of the explanation of ruffs at their hill reacting to a flying redshank as to a reeve (Portielje, 1931). The great variability at different times of the apparent meaningfulness of objects in the environment can only be understood in reference to the internal state. Yarrell (1843) mentions a buzzard which would rear chickens if allowed to hatch them from the egg, but devoured them if presented to her without a prior period of incubation. Sudden reversals of behaviour, due either to physiological or psychological causes, may occur. On the day when a pigeon's second brood hatches she attacks the fledglings of the former brood if they are in the cage with her (Craig, 1909); the expulsion of the young at the inception of a second brood is recorded of a variety of birds from the domestic fowl to the sea eagle (Goethe, 1937). A male kittiwake begins to feed his mate on the very day the first egg is laid (Perry, 1940). A pigeon, normally tolerant towards others on the ledge with him, drives them away when his mate returns (Mayr, 1935).

Internal changes or alterations in the environment by causing emotional changes and alterations in the specific energy available for certain behaviour-patterns raise some thresholds of response and lower others. Thus when a bird is outside its territory its behaviour will greatly differ from that within it.

Possibly still further differentiation within the releaser concept is required. Lack (1939a, 1940c) found that in the robin no single factor releases aggressive behaviour, for it will attack birds of other species, usually in flight, a mounted robin lacking a red breast, and a mere bundle of red feathers. Three different types of aggressive behaviour are respectively associated with these appearances; a small bird flying away provokes pursuit-flight, a stationary bird of robin shape elicits direct assault and a red breast stimulates threat-posturing. Thus there is a tendency, but only a tendency, for the three kinds of aggressive behaviour to be correlated each with its own releaser—and possibly also with different internal states. Lack suggests that the bird's original reaction is to a more general situation.

Greatly as Lorenz has advanced the understanding of bird behaviour there are other aspects of bird psychology which are still obscure. Such work as that of Hertz (1928, 1929) on pattern perception in birds indicates that there are phenomena which the releaser concept cannot interpret, and Howard's penetrating philosophical analysis of the bird mind in *A Waterhen's Worlds* (1940) suggests that the relationship between significant objects in the environment and a bird's response cannot be understood completely on the basis of releasers, innate perceptory patterns and internal states. He shows that the valence of objects for a bird and its reactions to them alter as it moves about its territory. He also raises more fundamental questions than Lorenz has dealt with when he considers the nature of perception in the bird. How far do interpretation, memory and expectation enter into perception? What precisely does 'recognition' of an object mean when we attribute it to a bird? To what extent is a bird's experience a single whole? Do we greatly falsify matters by separating mind-story from body-story? Our ignorance on such matters is still vast. But work such as Howard's warns us against too mechanistic an interpretation of bird behaviour. Analogies of films and television such as we have used represent the facts very inadequately, for the mind of a bird is not a blank screen awaiting impressions from outside, neither is it an elaborate machine. The bird's world is comparable to a dynamic system of ever-changing relationships. In the words of Hippocrates: 'Everything is in whole-part relationship, and, part by part, the parts in each part operate according to their function.'

CHAPTER XIX

THE RELATIONSHIP BETWEEN THE PHYSIO-LOGICAL AND PSYCHOLOGICAL ASPECTS OF DISPLAY

The problem of achieving breeding condition at the optimum season— Internal rhythms—Experimental evidence of endocrine influence on be-haviour—Effect of light on sexual development—Phases in the reproductive cycle—Recognition of sexual phases and pair-formation—Stimulatory effect of display—Social stimulation—Imitation—Summary of the factors which exercise a favourable effect on the breeding cycle—Bearing of the known facts on the theory of sexual selection.

THIS survey of bird ceremonial and display has shown with cumulative force as it proceeded that racial survival requires that pairs of birds should achieve co-ordination of their sexual cycles at the biologically 'right' or optimum time of year. Setting aside any consideration of the difficulties involved in the migratory journey, occupying a territory and securing a mate, let us consider some of the problems which be-set breeding birds in temperate regions. In a comparatively short favourable period they must build nests and rear young amidst fluctuating and frequently inclement weather conditions. Nest-building in itself may constitute a formidable task. A pair of nuthatches has been known to carry 6695 fragments to the nesting hole from birch and pine trees, of which the nearest were forty-five and seventy-five yards away, respectively (E. P., 1941). The temperature of some nestlings, such as the American house wren, varies with the atmosphere during the first few days of life (Kendeigh and Baldwin, 1928) so that they may die after an hour's exposure. Moreover, at the stage when it is most important that squabs should be kept warm their voracity is great. Young starlings and crows in their first few days eat half their weight of food daily, and later, a quantity almost equal to their own weight. For many species the problem of providing both food and warmth in wet, cold seasons is considerable. But disaster probably befalls eggs and young indirectly as often as directly at such times, for a fall in temperature may retard or inhibit the breeding rhythm of the parents. In exceptionally cold weather a yellow bunting will brood the young less than usual (Howard, 1929). The pattern of reactions becomes disorganised and thus we find incomplete clutches, infertile eggs and deserted young when the season is severe. In the tropics breeding may be initiated and carried on successfully over a long period—sometimes there is apparently no season of the year which is unfavourable—and few species suffer from food shortage; but in the temperate zone if young are to be reared it is essential that they should hatch when food is abundant. This end can only be attained by a prior co-ordination of the organisms with their environment and the female's sexual rhythm with the male's.

In considering the relationship of the factors which are operative in bringing about the synchronisation of the male and female sexual cycles and successful reproduction, it will be convenient to think of them in three groups—physiological, environmental and psychological—although any such classification must be somewhat artificial.[1] As three athletes of unequal degrees of strength and skill applying their force at somewhat different angles to a push-ball may yet direct it to its goal, so

[1] As an example of the inter-relationship of physiological, environmental and psychological factors in another connexion the regurgitation of food for mate or young by the black-headed gull may be mentioned. It has been shown by X-ray examination that motility in the stomach during cold weather may not begin for an hour, but that in warm weather it commences in a few minutes (Harris, 1937). Psychological stimulation comes from the importunate partner or chick.

these three groups of factors, influencing one another yet functioning to one end, are employed by, and operate upon, the organism in the struggle to reproduce its kind. The foundations of behaviour must be sought in the physiological processes of the animal, but they are what they are because of the environment in which they develop and operate. The influence of external factors may be direct or mediated psychologically; but when we speak of psychological factors we are reminded that no organism is merely passive or receptive to its environment but is capable of reaction in order to change it. A very complex series of reciprocal influences is involved in bird behaviour, and in studying these problems we are wandering amongst foothills which lead us within view of the unexplored peaks of the relationship of body and mind.

Various cycles may be discriminated within the life cycle of a bird (Whitman, 1919; Craig, 1909). Most of them may be considered as organised within the reproductive cycle and all of them may be classified in relationship to it, whether they be the spectacular passage movements or minor rhythms such as the brooding cycle. The degree of flexibility may vary, but sometimes the cycle operates in a mechanical and rigid way. In most species of dove the female broods from the evening until the morning of the next day, the male during the rest of the time. When a cat killed a domestic pigeon her mate kept strictly to his schedule and roosted in his usual place at night beside the nest—with the result that the young died. Nevertheless, the cock started to brood them at the customary hour of ten o'clock next morning (Lorenz, 1935).

There can be no doubt that birds possess an internal regulative mechanism or rhythm, but it is very difficult to prove experimentally that this is the case owing to the problems involved in devising means to eliminate or assess the effects of exteroceptive factors. Chapman (1928) noticed that Wagler's oropendola, Natterer's chatterer and a black-throated hum-ming-bird inhabiting a tree together in the Panama Canal Zone started to nest within a few days of the same date in successive years, although temperature remained remark-ably constant and the ending of the rainy season varied each year by several weeks. Other observations of a similar kind have been recorded by Moreau (1936) and Baker (1938). In the very equable climate of Hog Harbour, Espiritu Santo, New Hebrides, Baker and Bird (1936) found that all females of the insectivorous bat *Miniopterus australis* conceive at the same time within a few days. But facts such as these, which might appear at first glance to provide evidence of an internal rhythm, must not be taken at their face value. The regularity may be due to other factors. Baker himself considers an internal rhythm with an exact annual periodicity to be inconceivable (Baker and Baker, 1936).

None the less, there is evidence of an internal rhythm in various vertebrates. Ground squirrels (*Spermophilus tridecem-lineatus*) subjected to continual darkness developed sexually in spring in the ordinary way (Johnson and Gann, 1933). Hill and Parkes (1934) and Bissonnette (1933) found that the breeding season of ferrets subjected to reduced daily lighting or blinded was only slightly delayed. Bullough (1941) dis-covered that the gonads of minnows (*Phoxinus laevis*) kept in darkness developed fully, though tardily. He concludes that 'in some vertebrate animals, including the minnow, there exists an inherent reproductive rhythm, possibly of a more or less vague type, which in normal circumstances is reinforced and rendered more precise in its time of action by the effects of the seasonal variations of the external environment. In the absence of such seasonal variations the internal rhythm is strong enough to cause normal sexual development, although it is also weak enough to be overcome by the abnormal presence of spring conditions in winter'. As he points out (Personal communication), although there appears to be no experimental proof of internal rhythms so far as birds are concerned there is evidence of other kinds in favour of it. Living in the same environment British and Continental starlings are stimulated by increasing light in different degrees. The British birds, and not the Continental, become sexually active in autumn when light is decreasing.

We shall shortly return to the question of the influence of light on the breeding cycle, but it is desirable to mention here some of the evidence showing how potent are the effects on behaviour of the endocrine secretions and particularly of the male hormones. Experimental work on this subject has already been mentioned and it must suffice to give a concise summary indicating the direct influence of the internal secretions on the more conspicuous types of bird activity. Modifications in such secondary sexual characters as plumage are dealt with in the literature cited, but a discussion of such matters would occupy more space than is available here.

Migration. Schildmacher (1933) found that he could prevent redstarts migrating in autumn by injections of female sex hormone. Increasingly large doses were necessary as autumn advanced. Castration, which is equivalent to very acute regression of the gonads, does not prevent the autumnal migrations of the black-headed gull—indicating that this migration is a negative phenomenon (Putzig, 1937). Experiments by Hann (1939) in which castrated birds continued their migration and a red-eyed towhee returned after two years, and by Putzig (1939) proving that castration does not prevent the northern migration of hooded crows, show that further investigation is necessary before the nature of gonadal influence on the migratory impulse can be determined. The British starling, whose gonads do not regress as much as those of the strongly migratory Continental starling, in contradistinction to it only leaves the occupied locality during its first summer and autumn when the gonads show maximum involution (Bullough, in the press).

Territory. Pairs of British starlings hold territory in autumn under the influence of male sex hormone (Bullough and Carrick, 1940; Morley,

1941; Bullough, in the press). Tinbergen (1939 b) points out that in the same week as Eskimo dogs first copulate they become territory-conscious. A male coot, maturing early, will try to pen the whole flock into a corner of a pond, but if the pond freezes and a pair are forced out of their territory by ice they lose their territorial aggressiveness, showing that not only internal state but also a spatial field of reference is involved (Huxley, 1934; Howard, 1935).

Song. A female canary injected with male hormone sings like a male (Leonard, 1939). It is known that oestrogenic hormones may be produced by the testis and androgenic hormones from the ovary. Male hormone generated by the female starling alters the bird's bill in autumn to a bright yellow (Witschi and Miller, 1938). The female British starling, unlike the male, sings rarely in spring, but does so in autumn under the influence of male hormone. A correlation between the holding of winter territory, female song in winter and the production by the ovaries of male hormone may be found in birds with winter territories such as the robin, loggerhead shrike (Miller, 1931), mockingbird (Michener and Michener, 1935) and the wren-tit (Erickson, 1938). Küchler (1935) has shown that the robin's thyroid gland has two periods of activity, one in spring and another in autumn, and that in the yellow bunting the thyroid is more active in autumn than in spring. It is significant that in the one species there are two seasons of song and that in the other there is more autumnal than spring song. Injections of thyroid extracts cause the domestic cock to become broody (Ceni, 1927).

Dominance. The administration of a form of male hormone changes birds low in the social hierarchy to positions of dominance (Allee and Collias, 1938; Allee, Collias and Lutherman, 1939; Shoemaker, 1939 b; Bennett, 1939, 1940).

Display. Male chicks, 9 to 13 days old, injected with hebin (extract of the anterior pituitary) a few days after treatment crow and tread like older cockerels (Domm and Van Dyke, 1932). Noble and Wurm (1940 a) have demonstrated that injections of testosterone propionate radically alter the behaviour of black-crowned night herons. Month-old chicks under its influence behave like adults, occupying territories, building nests, exhibiting all adult display ceremonies, copulating and eventually brooding. In these, as well as in other birds on which similar experiments have been conducted changes took place in voice and adornments.

Such experimental results as these suggest that many different types of untimely behaviour will be found to be associated with excessive or precocious activity of the endocrine system—autumnal communal displays of ducks, blackcock, ruff, golden plover and other birds, territory defence and song in the female robin, the northward movement of newly fledged terns, premature and belated nest-building activities, and so forth. The rearing of more than one brood, sportive activities and a prolonged season of display may all be shown to be correlated with specific glandular activity. But while the discovery of these facts places the interpretation of bird behaviour on a sounder basis, he would be a superficial thinker who assumed that they lend themselves to a mechanistic interpretation of life. Territorial behaviour, for example, has psychological implications and spatial reference. Putting it crudely, before it became a glandular affair it was psychological, a matter of the relationship between perceiving birds and other birds and objects in the perceived environment.

Men have realised from immemorial times that the reproductive cycle is regulated in various ways by seasonal changes. There is now a great deal of experimental evidence proving that light is one of the most important external factors in furthering the onset of the breeding condition. It has long been the custom in Japan to expose pet birds to artificial light for some hours after sunset in order to bring them into song in January. Experiment has shown that testis progression and increase of song are concomitant (Miyazaki, 1934). Rowan (1931) exposed junco finches and Canadian crows during winter to ordinary electric lighting and found that the gonads increased in size as they do in spring. The crows, when liberated, went northwards fulfilling the spring migratory impulse. By artificially increasing the light hours of early spring Marshall and Walton (1936) succeeded in inducing ducks to display prematurely. Recrudescence of the gonads accompanies all such precocious behaviour. A great deal of experimentation, notably by

Bissonnette and Benoit,[1] has not yet elucidated the precise means by which this stimulation is effected. Intensity, wave-length and method of light-increase all apparently may play a part (Bissonnette, 1933), and length of day is known not to be the sole controlling factor; it is, however, clear that a certain minimum day-length must be reached for spermatogenesis to take place. Intensity of illumination will not compensate for the non-attainment of this duration-threshold (Burger, 1940). Baker (1939) has drawn attention to the fact that some Ceylon birds have different breeding seasons on the two sides of the island. There is good evidence that the eye, acting as a light receptor, stimulates the pituitary gland by nervous channels and that the activity of the pituitary controls the internal secretion of the testis. Experimental results which might seem at variance with this can probably all be explained by the operation of internal rhythms. The effects of the more active metabolism consequent upon the additional activity due to increased duration of lighting are difficult to assess. Some of the evidence brought forward by Rowan (1937, 1938) in this connexion has been found to be unsound. The London starlings whose enlarged gonads he attributed to the activity induced by artificial lighting and disturbance due to the noise of traffic were British birds whose gonads increase in size in advance of those of Continental birds (Bullough, in the press).

It is known that the ultra-violet light of daylight promotes egg-production (Whetham, 1933) and its abundance at certain seasons in the Arctic and Antarctic may have an effective influence on the breeding cycle of the birds. Marshall and Bowden's (1936) proof that ultra-violet light and intensity of illumination affect the onset of the cycle offers an explanation of twice-yearly breeding in tropical birds. The two breeding

[1] For surveys and bibliographies of this work cf. Marshall (1936, 1942) and Rowan (1938).

seasons of Ceylon birds synchronise with the two periods when
the intensity of illumination is greatest (Baker, 1937). It is
probable that ultra-violet light plays a part also in regulating the
breeding cycle of fruit bats (Baker and Baker, 1936). Rowan
(1938) thinks that different principles may be involved in the
action of visible and ultra-violet light. Birds such as the emperor penguin and the kea which breed in mid-winter present
a special problem. Possibly there is a time-lag in the response of these birds to the increase of light earlier in the year
or, alternatively, like the brook trout (*Salvelinus fontinalis*) they
may respond to decreasing light (Hoover and Hubbard, 1937)
just as certain flowers come into bloom as days become shorter
(Bissonnette, 1936 *a*).

Evidence has been furnished by Wynne-Edwards (1930) indicating that moonlight may affect ovulation in the nightjar. If
this be confirmed, it should be considered whether moonlight
may also be significant in the breeding cycles of the petrels.
Leach's fork-tailed petrel has a particularly energetic, nocturnal,
communal flight (Ainslie and Atkinson, 1937). Possibly the
greater activity of the nightjar in moonlight, when it can hunt
more successfully, influences the metabolism and so promotes
ovulation. It is conceivable that the sexual flights of birds have
such a function.

While it is beyond question that light is of much importance
in regulating the sexual rhythm and synchronising it with the
optimum breeding season, other factors exert a powerful influence in some species. Thus the fine adjustment is given to
the mechanisms involved. The adaptation of an organism to its
environment includes the evolution of the susceptibility to
particular influences which bring it regularly into breeding
condition.

The gonads of birds wintering in East Africa remain small in
spite of many hours sunlight, indicating that light is only one of

many factors governing sexual development; but immediately before their departure for the north they are found to be enlarging (Rowan and Batrawi, 1939). However, research by Stimmelmayr (1932), Wolfson (1940) and others suggests that the migratory impulse is controlled by more subtle factors than increase of light and development of the gonads. The pituitary may exercise a dominating influence, but migration is a response by the whole organism to many stimuli from within and without.

The development of the gonads is contemporaneous with the seeking of a territory. The general sequence of behaviour is, then, the acquisition of a territory, pairing-up (but some birds arrive at the breeding site already paired) preceded and followed by a varying amount of display, coition, nest-construction (with display and coition continuing), ovulation and egg-laying on the part of the female, and then incubation and tending the young by one or both of the pair.

It is a commonplace of agricultural practice and animal husbandry that food supply affects the breeding cycle (Marshall, 1922) and dietary anoestrus is a well-known phenomenon in mammals. There is evidence of similar phenomena in birds. If starlings are kept on a low diet, their sexual rhythm does not respond to light as it does when they are well fed (Bissonnette, 1933). The gonads of sparrows supplied with plenty of food develop normally in the absence of increasing amounts of light or other stimulating factors (J. C. Perry, 1938). During vole plagues short-eared owls increase their clutches from four to eight up to as many as fourteen, and though normally single-brooded may have two clutches. What factors cause these birds to leave their winter haunts in Ireland and settle in pairs where voles are common we do not know, but the possibility that abundant food-supply may stimulate the sexual rhythm visually should not be ruled out *a priori*.

Certain it is that weather conditions affect the breeding cycle. Their direct influence is shown in the reduction of the tempo of nest-building when the weather is inclement (van Someren, 1933). Whitman (1919) found that white ring doves did not remove the egg-shells from the nest in cold weather and that when temperature was low the male neglected to brood. Ovulation and egg-laying of Arctic terns are retarded if the season in high latitudes is late and involution of the gonads may occur in very late seasons (Lack, 1933). The influence of the weather is apparently mediated visually in such cases, for when the snow clears the birds nest. Certain species near Lake Nyasa nest over burnt veldt and where the first local fires occur the nests are found crowded together before there are any in the main areas (Wilkes, 1928). Craig (1913) found that ovulation was delayed if birds were disturbed near their completed nest. Even the presence of a strange object or bird may have this effect. Non-breeding is a well-known phenomenon amongst birds in the Arctic, such as the long-tailed skua, snowy owl and greenland falcon. In some cases it may be correlated with the scarcity of lemmings, but the causes operative amongst king eiders and red-throated divers are more obscure (Bird and Bird, 1940). Blizzards which prevent the courtship flights of knot, sanderling and dunlin may have an important negative influence.

In temperate climates the most favourable conditions for breeding are warm, fairly calm and dry weather. Excessive drought or extreme weather of any kind usually exerts an inhibitory influence, but in every region exceptions may be found, such as the crossbill in England, the nutcracker in the Alps, and kea in Australia which breed during seasons of storm and low temperature. In the tropics the inception of breeding often follows close upon the breaking of the rains and the consequent abundant food supply (Baker, 1938); conversely, when the rains fail the birds may not breed (Belcher, 1930).

Up to this point our discussion has been concerned with the

PLATE XXI

page 236

LYRE BIRD
Male singing and displaying in his court.

PLATE XXII

page 156

CORPORATE FLIGHT OF FLAMINGO COLONY

GANNETRY

Note the uniform size of the nesting territories.

factors, internal and environmental, which contribute to bring birds into breeding condition, but so far we have confined ourselves to a consideration of the bird apart from its relationship to other birds. If this has been somewhat like playing Hamlet without the Prince, it has been expedient in order to simplify the analysis of breeding behaviour, but we must now consider the breeding cycle as the co-ordination of two birds' lives in order that they may procreate their kind.

According to Howard (1929) there are four phases in the reproductive cycle of a pair of birds, but the first of these has reference only to the male. During it he claims a territory, sings, is aggressive and shows a greater or lesser disposition to build. The other phases Howard describes by the female's condition as pro-oestrum, oestrus and another phase which is not comparable with anything in the sexual cycle of a female mammal. Pro-oestrum is characterised in pipits, buntings, warblers and other species by sexual flight in which the male chases the female and both thereby sustain stimulation; the female is sexually excited, but her responses are inadequate. She fidgets with nesting material, postures ineffectively, but does not build or copulate. The frequency and intensity of posturing vary from day to day. This phase may be defined as the period from the first manifestation of sexual activity to the first successful copulation. After this pre-nuptial period the secondary physiological control which Howard postulates in regard to the female as distinct from the male is lifted and the inception of the oestrous phase is indicated by coition and nest-construction. It is difficult to define this phase in any precise fashion other than to call it the period during which successful copulation takes place. The third and final phase, according to Howard, sees the completion of the nest, laying of eggs, waning of the sex-attraction of one bird for the other, and the tending of the young. But no behaviour specific enough to give exact definition to the period following oestrus seems to be observable.

It is somewhat doubtful whether terminology taken from mammalian physiology is usefully applicable to birds or whether their ovarian processes are sufficiently comparable with the sexual changes of female mammals to justify the use of these terms, but they may have to be employed until more appropriate names for avian reproductive phases are available. As female birds may copulate long before ovulation takes place and males attempt copulation before sperms are available, the observer who defines phases in terms of behaviour is liable to become at cross purposes in his terminology with the physiologist who relies on the results of dissections.

The evidence already quoted has shown the close correlation between internal states and the initiation of certain patterns of behaviour. In female buntings and warblers coition and nest-building are coincident and in robins the first copulation and the first courtship-feeding occur within a day of each other (Lack, 1939a). But so many exceptions to most of the 'rules' are known and the life-histories of so few birds have been investigated in detail that generalisations purporting to rest upon correlations between internal state and behaviour true of birds as a whole are apt to be precarious.

So far is nest-building from being concomitant with coition that, as we have seen, immature birds may show nest-building behaviour and in exceptional instances supernumeraries may aid mated pairs in building or feeding young. Nest-construction of a kind may continue all through and even beyond the breeding period, as in the case of the gannet mentioned earlier. Howard (1929) quotes Whitman (1919) to the effect that incubation and sexual activity are mutually exclusive and antagonistic phenomena, and that coition ceases when incubation begins. He says that he has never known an exception to this rule. But copulation during incubation has been recorded of many birds, including the black-headed gull (Kirkman, 1937), rook (Yeates, 1934), corn bunting (Niethammer, 1937–8), nightjar (Lack,

1932), Arctic tern (Bullough, 1942), some hawks (Murphy, 1936), great crested grebe and the common sparrow. Tinbergen mentions a snow bunting inviting coition before the young had left the nest. The wandering albatross is as ardent in copulating after the young have hatched as before incubation begins (Wilkins, 1923). Lack (1932) records a form of sexual flight when a pair of nightjars were incubating. It is true, as Howard points out, that a female stonechat gives up feeding the chicks of the first brood when she enters on the cycle of the second brood, yet it is also true that the white-eared humming-bird feeds full-fledged young in the intervals of incubating the second clutch (Skutch, 1935). An instance is on record of a robin having four broods, the last three overlapping (Osmaston, 1934). Indeed, overlapping broods are not uncommon as the records collected by Tinbergen (1939a) and Fisher (1939) show. They have been recorded as occurring regularly in the blackbird, great tit, nightjar, various pigeons, ringed plover, waterhen and song sparrow; and also of the kingfisher, wagtail, redstart, whinchat, pallid shrike, red-backed shrike, robin, goldcrest and reed warbler. In some of these the male looks after the young of the first brood while the female busies herself with the new nest, but ringed plovers take it in turn to brood the eggs and take charge of the chicks (Koehler and Zagarus, 1937). I have seen a willow wren feeding fledged young while on migration.

Howard's opinion expressed in *An Introduction to the Study of Bird Behaviour* that the male is ready to build, incubate and care for the young from the moment he takes up territory, and more boldly enunciated in *The Nature of a Bird's World*—that, after occupying a territory he is 'always ready to complete the sexual act, to build and to brood' whereas the female is not, requires qualification. The truth enshrined within it is that in many species the male is ready and desirous to inseminate the female for a longer period than she is able to satisfy him. Tinbergen

(1939a) found that the male snow bunting seemed to be ready for copulation before the female reached the oestrous phase and remained sexually potent after the female's oestrus had terminated. Roberts (1940a) has shown that in the male gentoo penguin there is a period of about six weeks during which copulation may be successful, but he was unable to trace any correlation between the behaviour of the females and the size of the most advanced follicle in a series of ovaries. Bullough notices that sperm is present in the testes of the male starling in early March, but is not available until mid-April. Behaviourally the female is as active as the male. The period during which successful copulation may take place is about a fortnight, though the female's oestrus may be prolonged if insemination has not been attained. Allen (1934) believes that male birds are similar to the females in having a short, definite mating period and that it varies with species, individuals and especially with age; but Gould's manakin is sexually active for at least four months and possibly eight in the year (Chapman, 1935). As has been mentioned copulatory behaviour may take place when the males are unable to inseminate the females and in some such cases we may be dealing with a close succession of oestrous phases rather than one continuous period.

In not a few species the males take a succession of females for successive broods. There is evidence that the desertion of the mate is due to her inability to respond to renewed sexual advances, but if the male is unresponsive to the female she may leave him to find an active partner. This happened in the case of a snow bunting whose mate was occupied in feeding the young of the first brood and therefore was unable to copulate (Tinbergen, 1939a). Here is an instance of one activity—or rather the internal state concomitant with it—precluding another. Through the researches of Riddle (1935) and his co-workers (Riddle and Dykshorn, 1932; Riddle, Bates and Lahr,

1935, 1937) we have some insight into the physiological foundations of such behaviour. They have shown that prolactin, which is secreted by the anterior lobe of the pituitary and is apparently a hormone, induces broodiness but also causes regression of the testis and ovary, and indirectly inhibits ovulation in the female. Here we have a clue as to the nature of what Howard called secondary physiological control, but the evidence, both observational and physiological, indicates that though such 'control' may operate in differing degrees in the two sexes it is not confined, as he believed, to the female.

Howard (1929) expresses the difference between the male and female in these respects in the passage which follows. After describing how a bird occupies a territory he says:

Next, a male is influenced by other males; and lastly by a female. But in a female the order is reversed. There is nothing to show that the physical world exercises the sort of influence upon her that it does upon him. She receives her first influence from a male; stimulates his sexual nature and seems to enjoy the consequences; becomes aggressive and toys with nesting material. Yet, though sexually excited, she cannot complete the pairing act; though she toys with nesting material she cannot build; but she can fight as determinedly as her mate. The difference between them lies, then, not in the pattern of reactions which, except in one respect—the physical relatedness—is alike, but in the relative responsiveness of the parts.

This is what Howard means by a secondary physiological control. The cessation of the control is the inception of the oestrous phase according to Howard's definition. He maintains that if the females did not suffer this prior sexual condition but when they paired were in a state of full sexual excitability —instead of having a period of from one to five weeks in which to find mates—the struggle for partners would involve sudden excessive competition when the females arrived and the young would be brought into existence when there was neither food nor warmth available for them. The pre-nuptial period may vary greatly as between individual pairs and different species.

In the robin it may be more than fifteen weeks (Lack, 1939 a), in the marsh warbler three days (Howard, 1929). Yellow bunting pairs formed on 17 February and 6 March respectively hatch on the same date (Howard, 1929). This is the usual state of affairs; birds which pair-up late have a shorter pre-nuptial period than those paired earlier and thus eggs of a given species tend to be laid at approximately the same time each year. Tinbergen (1939a) records a female snow bunting which copulated with a mate for her second brood almost immediately after she joined him.

The conclusion which may be based on our as yet imperfect knowledge of these matters is that in many species of bird breeding in temperate climates the female appears to be, as Howard has maintained, under a stricter physiological control than the male; but whether this control is primarily due to internal rhythm or susceptibility to external conditions, such as temperature, light and the like, we can only speculate. For all we know the long pre-nuptial period in some species may be due as much, or even more, to the tardy maturation of the male. It would seem that a system has been evolved which gives the females a fairly long but variable period to find mates and achieve insemination, intensifies sexual susceptibility at the optimum time for breeding to begin, avoids too violent and sudden competition, gives scope for the male by his display to excite the female further, and yet enables her, if the fine adjustments to her breeding condition available from association with and stimulation by the male are lacking through tardy pairing-up, to dispense in some measure with these, or rather to achieve them largely through her own internal development. But here we are on speculative ground.

So far as the male is concerned the occupation of a territory is, as a general rule, a necessary preliminary to breeding. Usually the nest is in his holding, but this is not invariably the case, as our study of lek birds showed. In a few instances, such as Wagler's oropendola, it is the female who secures and defends something approximating to a territory—the twigs on which the nest is built. Pairing-up in the male's territory may occur, as we have seen, months before copulation first takes place as in the robin (Lack, 1939 a) and the British starling (Morley, 1941), in both of which species the female takes the initiative. Female migrants returning north to breed arrive some days later than the males and pair-up in their territories. Occupation of a territory is the first step towards successful breeding by the male whose internal condition has matured, pairing-up the first step by the female. Thus pairing-up marks the initiation of the process of synchronisation of male and female. Birds are able, when we are not, to recognise the sexual condition of a potential partner. Consider the behaviour of a whitethroat which according to Howard (1929) arrived on 28 April, established a territory and behaved in the usual way, singing, fluttering up into the air and chasing other birds. He was visited in succession by five females. In each case, when he perceived any one of the first four birds he flew down from a chestnut tree, frolicked in the air and then went about his business. But when the fifth appeared he acted quite differently, dancing over her at intervals, pursuing her in sexual flight at the same time uttering a hitherto unheard note and singing but little. They settled down together and bred. Howard's deduction is that if a female elicits adequate sexual reaction in a male she stays, but if she fails she leaves. Ponting (1921) describes how an Adélie penguin, newly arrived from the sea, inspected and rejected three possible partners and then courted a fourth. When a yellow bunting's mate comes into full breeding condition neighbouring males, hitherto satisfied with their own mates, become excited and endeavour to get at her (Howard, 1929). Birds of the British race of starling occupying nesting territories realise in some way that Continental starlings are at a low stage of their sexual cycle and tolerate them near their holes, whereas trespassing British birds are driven away with fury (Bullough, in the press). We do not know how a bird recognises another as a suitable mate just as Chapman (1935) is unable to suggest how a nesting female manakin is perceived by the males not to be at that stage in which their favours are acceptable. A bird, male or female, in a certain phase has the capacity to recognise that another bird is in an appropriate corresponding physiological condition.

Pair-formation having taken place with or without a certain amount of display, the posturing of the paired birds has the effect of establishing in-phase correlation between them. Chapman (1935) says of Gould's manakin, 'Whatever be the sexual condition of the female she apparently must be courted before she will receive the male.' Similarly Bristowe (1929), writing of certain spiders, thinks that without prior courtship display it is impossible for the female to copulate. Selous (1909–10) reached the conclusion that greyhens come to the lek for the definite purpose of being aroused sexually, and if the stimulation is not sufficient they depart without coition having taken place.[1]

The prime importance of display stimulation is shown in another way. Dewar (1908) noticed that the reason why the peahens in Lahore Zoo showed a decided preference for white cocks was not because of the purity of their plumage but because they continually made sexual displays, eroticism being more pronounced in them than in normal birds. It is not that the females prefer albinos but that they, by their activity in display, are able more effectively to arouse sexual response in the females. Amongst spiders we find a similar state of affairs. A species of the family Attidae, *Astia vittata*, is dimorphic, having forms which exhibit different sexual dances of which one is much livelier than the other. When the two types of male with

---

[1] Without sex play the reproductive cycle of certain frogs, toads, newts, lizards and fish apparently cannot be completed (Marshall, 1942).

their characteristic dances come into competition with each
other the female prefers the more active performer (Peckham,
1889, 1890).

While there is no doubt that the sexual cycle of both sexes is
under the control of the internal rhythm and environmental
conditions, the extent to which one sex needs to be aroused by
the other sex is still only vaguely known. Howard's opinion
that the male requires stimulation less than the female and that
he may not need it at all in some instances seems to be generally
true of the birds which he studied, but species and individuals
vary. Tinbergen (1939 a) found insufficient evidence in the case
of the snow bunting to warrant the conclusion that the male
initiates more often than the female. He believes that the different way in which she indicates her desire, together with her
long pre-nuptial period, have given rise to the mistaken idea
that the male is the more ardent. In red-necked phalaropes each
sex has its turn of eagerness (Tinbergen, 1935). Apparently
there are lesser rhythms within the oestrous phase and the function of connubial display is to synchronise these.

Stimulation advances step by step, as we know from
Carpenter's (1933) researches on the pigeon. There are ample
grounds for believing that what he found to be true of this
species in respect of rising thresholds of sexual ardour may be
applied to the interpretation of the sexual display of most birds.[1]
The first and most generalised secondary sexual activity is the
preening of one bird by another; this may be observed when the
squabs are only two or three weeks old and occurs amongst
segregated females, but is most frequent in sexual situations.
There may also be noticed what Carpenter calls 'wing-plucking'
—a somewhat similar gesture to that described earlier in referring to the display of wood pigeons, albatrosses and cranes. The

---

[1] In black-headed gulls Rothschild (1941) records that the display is
built up by the addition of successive components. At the height of the
breeding season her birds displayed in their sleep.

bird throws its head backwards and pokes its beak under one of
its wings. This activity has not such a definite place in the series
of display actions as the other postures described here. Next in
order of stimulatory effect, and often also in time, comes billing.
Male and female clasp bills and move their heads with a pumping motion as if they were feeding a chick. Often liquid exudes
from the interlocked bills. The duration of this courtship
feeding phase varies from one to thirty seconds. The activity
styled 'charging' occurs frequently in the sequence of posturing,
but it represents a higher stage of emotional stimulation than
billing or preening. The pigeon ruffles its neck and back feathers,
lowers and slightly spreads wings and tail, tilts its body forward
at a sharp angle and advances in spurts, often accompanying the
display with muffled coos. I have noticed that the female meanwhile 'false-feeds' nervously. This posturing indicates a high
threshold of stimulation and is very provocative. With modifications it is characteristic of many birds. The sequence may
vary somewhat, but the hierarchy of intensity of feeling and
provocation seems to be—preening, billing, charging. Wing-
plucking often precedes billing and may be on a slightly higher
threshold than preening. Preening may recur after coition as
emotion subsides. Carpenter proved the intimate connexion
between the sex organs and psychological activities, for he found
that castration tended to disintegrate the order of display and
the whole pattern of behaviour. His conclusion gives additional
support to that view of the function of display to which this
whole discussion has led: 'It seems evident that the biological
function of the secondary sex activities is that of establishing the
essential, in-phase, synchronised relationship of excitation and
readiness necessary for the occurrence of primary sex activity
and other reproductive behaviour.'

The close inter-connexion of psychological and physiological
factors is shown in this writer's suggestion 'that the cyclic
character of the reproductive activities of the male pigeon is

determined, to a large degree, by the changing behaviour of his female mate to which he responds, and to the resultants of the male's and female's social behaviour, i.e. nest, eggs and young'.

As previous pages have already shown, it is by no means only from a bird's mate or potential mate that it receives sexual stimulation. The presence and activities of other birds affect the breeding cycle in a variety of ways:

1. We have seen that the lack of a suitable nesting site may have an inhibitory effect upon ovulation and that there is therefore a strong presumption that the presence of a suitable site and appropriate material may have a stimulatory effect. Apparently the presence of other birds engaged in nesting activities is also stimulatory. Penguins seem to be subject to such influences (Roberts, 1940a) and it is possible that the ceremonial presentation of material for the nest by herons and other birds is of significance in this connexion (Verwey, 1930). Certainly 'courtship nesting' in terns, lapwings and red-necked phalaropes has a stimulating effect (Tinbergen, 1935).

2. It is possible that ovulation is induced in the cuckoo by the sight of the proposed victim building its nest (Chance, 1922, 1940). An instance is on record of apparent inhibition of ovulation in a sarus crane when extra eggs were supplied (Lack, 1933).

3. When two female cranes or swans live in very close association egg-laying may occur (Tavistock, 1936). An isolated female pigeon will not lay, but if a female companion is introduced both may start producing eggs. A mirror in the cage or even stroking the bird's head may cause ovulation. In connexion with this Harper (1904) wrote:

The function of ovulation is in a state of tension, so to speak, that requires only a slight stimulus, 'mental' apparently in this case, to set the mechanism working. At any rate it is impossible to regard the presence of the sperm in the oviduct as an essential stimulus to ovulation, although it may have an important influence in the normal case. Our attention is directed to the various and complex instincts of the male which come under the head of courtship both before and after mating is effected, as furnishing a part of the stimulus to the reproductive organs.

It has been proved by Craig (1911, 1913) and Matthews (1939) that the stimulus involved when two birds are thus associated is visual and not tactile, olfactory or auditory. A sense of smell has, however, been proved to exist in the blue tit, blackbird, robin, whitethroat, greenfinch and mallard (Zahn, 1933), so that the possibility of some olfactory stimulation in some instances should not be regarded as out of the question.

4. So effective may be the provocative behaviour of another bird that a castrated pigeon can be aroused to reciprocate the solicitous salutations of the female and eventually to copulate with her, although no part of the motivation comes from the sexual secretions (Carpenter, 1933).[1] Analogous cases have been reported of the human species (Pratt, 1932).

5. Evidence from the study of birds displaying socially or breeding colonially supports the view that birds are stimulated sexually by being amongst their fellows, hearing their calls, and perceiving their display performances. They may even be excited sexually by the presence of other species. Thus lesser black-backed gulls nesting in a herring gullery laid at an early date contemporaneous with the earliest egg-laying of the species in its own colonies. The instances of green sandpipers displaying on seeing greenshanks copulating should also be taken into consideration.

6. It has been shown in a preceding chapter that a high position in the dominance hierarchy favours breeding, a lowly status militates against it. Success in fighting is of varying degrees of importance according to species, nesting habits and social organisation. The close connexion between territorialism and the social hierarchy has already been considered.

[1] Castrates which copulated freely showed much reduced brooding and nesting activity.

7. Self-stimulation operates in a degree which is difficult to assess. For example, the gentoo penguin indulges in solo display early in the breeding season. Some ornithologists, such as Whitman (1919) and Howard (1929) have regarded posturing generally as self-stimulating and held that it raises the level of intensity of the reaction. The analogy of a heated argument, suggested in Chapter VIII is a reasonably close approximation to the kind of stimulation which occurs. We know, for example, that successful threat display accompanies and probably promotes sexual vigour, but a position of subserviency in the social hierarchy inhibits it (Allen, 1934). Huxley (1923), referring to mutual ceremonies, styled them 'self-exhausting' in so far as they do not lead to copulation, but later (1938 b) admitted they 'may really be stimulative in regard to reproductive function'. A more suitable term is needed. Such mutual ceremonies, even though not prefatory to coition, may have a reciprocally stimulating and valuable connubial effect. There is so close a relationship between the concentration and discharge of emotional energy during display that any discussion as to the relative importance of one or the other is, in the present state of our knowledge, apt to be inconclusive. We may confidently believe that whatever functions display may have, either as a means of working up feeling or releasing it, all bird (and human) dancing, display and ceremonial operate in a regulative capacity, both stimulating and indirectly controlling the emotions. As Howard (1929) says: 'The oftener the female is stimulated the more susceptible she will become; and the oftener his reaction overflows into posturing, since he, too, derives pleasure from it, the more habitual his posturing will become.' On such lines the prolonged display season of various birds may be explained.

Any survey of the various ways in which birds stimulate each other would be incomplete without stress being laid on the significance of imitation both as indicating the susceptibility of one bird to another's influence and as one of the means by which stimulation operates. Many instances of the contagiousness of behaviour have been referred to in the course of the preceding pages and it is sufficiently clear that there is hardly any activity of birds which may not be aroused by the presence of other birds performing. The herring gull, for example, shows infectious behaviour in regard to visual and auditory display (Darling, 1938), preening (Goethe, 1937), courtship feeding (Goethe, 1937) and copulation (Richter, 1939). An instance of contagious copulation amongst avocets was mentioned earlier. As has been shown, even 'injury-feigning' may be contagious. Clearly this influence works powerfully amongst social birds and especially amongst those which display within sight of each other on a favourite arena. Taken in conjunction with Johnson's (1941) observations which suggest that northern guillemots in congested colonies benefit by the stimulation associated with squabbling it is easy to understand that the ceremonial of birds needing special stimulation would evolve in the direction of arena display.

Summarising the factors which are known to exercise a favourable influence on the breeding cycle prior to egg-laying we have the following:

Internal Factors

1. The state of the reproductive system, the growth of the gonads and the secretion of sex hormones.
2. General metabolism, including adequate food supply.

Environmental Factors

3. The effect of light.
4. Favourable weather.
5. The availability of a suitable nesting site and nesting material.

Psychological Factors
6. The possession of a territory.
7. The presence and display of a mate or potential mate.
8. Social stimulation.
9. High position in the social hierarchy.
10. Self-stimulation.

The overt elements mentioned in the previous chapter which contribute to the effectiveness of a bird's sexual or epigamic display, and also the threat or aposematic display, are as follows:

Vividness. This involves the possession of clear-cut releasers, whether morphological or behaviouristic.

Vigour. This manifests itself in the energy of posturing, erection of plumes, flight and other movements, intensity and assiduity of song and especially suddenness and surprise in display.

Repetition. There is ample proof of the cumulative potency of sustained and frequent display.

The subtle inter-relationship of these factors needs no emphasis. Environment, nervous system and endocrine system reacting upon each other regulate sexual periodicity and the sequence of stages in the breeding cycle. The environment conditions sexual periodicity and its successive phases are regulated by the combined integrative action of the nervous and endocrine systems (Marshall, 1936).

The inter-connexion of the factors in the two first groups has the broad effect of bringing the birds into breeding condition; the fine adjustments which achieve the final in-phase synchronisation are given by those in the third group. Posturing of one kind or another is associated with every item in this section. This fact reveals the essential nature of display.

Marshall (1936) has summarised the relationship between display, the nervous system and the condition of the sex organs thus:

It has been shown that the gonad-stimulating hormone of the pituitary will cause ovarian development and ovulation in birds, and that sexual posturing or even the mere association of two individuals will initiate nest building and ovulation. There is a presumption, therefore, that sexual posturing produces exteroceptive stimuli which act upon the anterior pituitary through the hypothalamus, and so effects the necessary synchronization between the sexual processes of the male and female birds. Herein then, in all probability, lies the biological or race-survival value of sexual display and of the adornment which in many species is taken advantage of to render the display more effective. Those birds which have brighter colours, more elaborate ornamentation, and a greater power of display must be supposed to possess a superior capacity for effecting by pituitary stimulation a close degree of physiological adjustment between the two sexes so as to bring about ovulation and the related processes at the most appropriate times....

The primary periodicity is a function of the gonad, the anterior pituitary acting as a regulator, and the internal rhythm is adjusted to the environment by the latter acting on the pituitary, partly or entirely, through the intermediation of the nervous system. The further fact, however, must not be overlooked, namely, that in the absence of the anterior pituitary the functions of the gonad fail, so that the pituitary, in common with the other endocrine organs, conditions the metabolic processes which are essential for reproduction.

It has not been possible in this book, with its predominantly psychological approach, to make more than incidental reference to the adornments associated with bird ceremonial and display, but we have seen that most display takes place after pair-formation, and therefore the theory of sexual selection as formulated by Darwin cannot account for the evolution of display adornments, except possibly in a few species such as blackcock, ruffs and some ducks. But Darwin (1871) was not far from the truth when he wrote of the female bird: 'It is not probable that she consciously deliberates; but she is most excited or attracted by the most beautiful, or melodious, or gallant males.' He appreciated the fact that the female's response was greatest to the male with the greatest capacity to stimulate her, and realised that in the evolutionary process the adornments of birds received modification through the psychological reactions which they induced in other birds. He was a pioneer in regarding psychical selection as an important process in evolution. Modern research has upheld him, particularly in emphasising the correlation between the senses and higher nervous centres of living creatures and the adornments and display behaviour. He was mistaken in his belief that unless the male's display induced the female to select him it was without meaning. The female does not choose the most beautiful male or the most 'artistic' performer as such, though there is evidence that Darwin was correct in attributing some aesthetic sense to female birds (Huxley, 1938 c); she responds to him who most effectually arouses her. But the process is reciprocal, for the male tends to reject an inadequate female or, what amounts to the same thing, such a bird does not elicit the appropriate response in the male. Forces are at work which select the fittest and it is still true to say that Nature eliminates the unfit. The fit, indeed, select themselves and one another.

In the course of the foregoing pages we have seen constant evidence of pattern, rhythm and adaptation in nature. There is perpetual striving to achieve that harmony between male, female and environment which is attainable by means of delicate adjustment to a multiplicity of influences. On this view the most successful organisms are those which most nearly become exquisite microcosms reproducing in miniature the synchronisation of parts and equipoise of the whole which we think of poetically as the harmony of the spheres.

Scientific Names

Order STRUTHIONIFORMES
Family STRUTHIONIDAE
Struthio camelus camelus Linnaeus — North African Ostrich
Struthio camelus australis Gurney — South African Ostrich

Order CASUARIIFORMES
Family CASUARIIDAE
Casuarius sp. — Cassowary
Family DROMICEIIDAE
Dromiceius n. hollandiae (Latham) — Emu

Order TINAMIFORMES
Family TINAMIDAE
Tinamus major castaneiceps Salvadori — Chestnut-headed Tinamou
Nothura maculosa (Temminck) — Spotted Tinamou

Order SPHENISCIFORMES
Family SPHENISCIDAE
Aptenodytes patagonica J. F. Miller — King Penguin
Aptenodytes forsteri G. R. Gray — Emperor Penguin
Pygoscelis papua (Forster) — Gentoo Penguin
Pygoscelis adeliae (Hombron & Jacquinot) — Adélie Penguin
Eudyptes crestatus (J. F. Miller) — Rockhopper Penguin

Order GAVIIFORMES
Family GAVIIDAE
Gavia stellata (Pontoppidan) — Red-throated Diver
Gavia arctica arctica (Linnaeus) — Black-throated Diver
Gavia immer immer (Brünnich) — Great Northern Diver or Loon

[1] The list of families follows the classification of Wetmore (1940) and the scientific names adopted are those of Peters (1931–40) to the end of the fourth volume, the last available.

Order COLYMBIFORMES
Family COLYMBIDAE
    Poliocephalus ruficollis ruficollis (Pallas)   Little Grebe or Dabchick
    Colymbus auritus Linnaeus   Horned Grebe
    Colymbus cristatus cristatus Linnaeus   Great Crested Grebe
    Aechmophorus occidentalis (Lawrence)   Western Grebe

Order PROCELLARIIFORMES
Family DIOMEDEIDAE
    Diomedea exulans Linnaeus   Wandering Albatross
    Diomedea irrorata Salvin   Galapagos Albatross
    Diomedea nigripes Audubon   Black-footed Albatross
    Diomedea immutabilis Rothschild   Laysan Albatross

Family PROCELLARIIDAE
    Fulmarus glacialis glacialis (Linnaeus)   Fulmar

Family HYDROBATIDAE
    Oceanites oceanicus (Kuhl)   Wilson's Petrel
    Oceanodroma leucorhoa leucorhoa (Vieillot)   Leach's Fork-tailed Petrel

Order PELECANIFORMES
Family PHAETHONTIDAE
    Phaethon lepturus catesbyi Brandt   White-tailed Tropicbird

Family PELECANIDAE
    Pelecanus onocrotalus Linnaeus   Asiatic White Pelican
    Pelecanus erythrorhynchos Gmelin   American White Pelican
    Pelecanus occidentalis Linnaeus   Brown Pelican

Family SULIDAE
    Morus bassanus (Linnaeus)   North Atlantic Gannet
    Sula nebouxii Milne-Edwards   Blue-footed Booby
    Sula dactylatra Lesson   Masked Booby
    Sula leucogaster (Boddaert)   Brown Booby

Family PHALACROCORACIDAE
    Phalacrocorax carbo carbo (Linnaeus)   North Atlantic Cormorant
    Phalacrocorax carbo sinensis (Shaw)   Southern Cormorant
    Phalacrocorax aristotelis aristotelis   Shag
    (Linnaeus)
    Nannopterum harrisi (Rothschild)   Galapagos Flightless Cor-
        morant

Family ANHINGIDAE
    Anhinga anhinga (Linnaeus)   American Snake-bird

Family FREGATIDAE
    Fregata magnificens Mathews   Magnificent Frigate-bird

Order CICONIIFORMES

Family ARDEIDAE

Ardea cinerea cinerea Linnaeus	Grey Heron
Butorides virescens virescens (Linnaeus)	Little Green Heron
Bubulcus ibis ibis (Linnaeus)	Buff-backed Heron
Hydranassa tricolor ruficollis (Gosse)	Louisiana Heron
Nycticorax nycticorax nycticorax (Linnaeus)	European Night Heron
Nycticorax nycticorax hoactli (Gmelin)	Black-crowned Night Heron
Nyctanassa violacea (Linnaeus)	Yellow-crowned Night Heron
Ixobrychus minutus minutus (Linnaeus)	Little Bittern
Ixobrychus involucris (Vieillot)	Variegated or Little Red Bittern
Botaurus stellaris stellaris (Linnaeus)	European Bittern
Botaurus lentiginosus (Montagu)	American Bittern

Family COCHLEARIIDAE

Cochlearius cochlearius (Linnaeus)	Boatbill

Family CICONIIDAE

Ciconia ciconia ciconia (Linnaeus)	White Stork
Ciconia nigra (Linnaeus)	Black Stork

Family THRESKIORNITHIDAE

Threskiornis melanocephala (Latham)	Black-headed Ibis
Theristicus caudatus (Boddaert)	Black-faced Ibis
Platalea leucorodia leucorodia Linnaeus	European Spoonbill

Order ANSERIFORMES

Family ANHIMIDAE

Chauna torquata (Oken)	Common Screamer
Chauna chavaria (Linnaeus)	Crested Screamer

Family ANATIDAE

Cygnus cygnus cygnus (Linnaeus)	Whooper Swan
Cygnus olor (Gmelin)	Mute Swan
Cygnus melancoriphus (Molina)	Black-necked Swan
Anser anser (Linnaeus)	Grey Lag Goose
Alopochen aegyptiaca (Linnaeus)	Egyptian Goose
Cairina moschata (Linnaeus)	Muscovy Duck
Tadorna tadorna (Linnaeus)	Sheldrake
Anas platyrhynchos platyrhynchos Linnaeus	Mallard
Anas crecca crecca Linnaeus	Teal
Anas eatoni eatoni (Sharpe)	Kerguelen Teal
Anas bahamensis Linnaeus	Bahama Pintail

Order ANSERIFORMES (*continued*)

Family ANATIDAE (*continued*)

Mareca sibilatrix (Poeppig)	Chiloe Wigeon
Chaulelasmus streperus (Linnaeus)	Gadwall
Spatula clypeata (Linnaeus)	Shoveler
Aix sponsa (Linnaeus)	Wood Duck
Somateria mollissima mollissima (Linnaeus)	Eider
Somateria spectabilis (Linnaeus)	King Eider
Mergellus albellus (Linnaeus)	Smew
Lophodytes cucullatus (Linnaeus)	Hooded Merganser
Mergus merganser merganser Linnaeus	Goosander
Mergus serrator Linnaeus	Red-breasted Merganser

Order FALCONIFORMES

Family CATHARTIDAE

Coragyps atratus foetens (Lichtenstein)	South American Black Vulture

Family ACCIPITRIDAE

Pernis apivorus apivorus (Linnaeus)	Honey Buzzard
Milvus milvus milvus (Linnaeus)	Red Kite
Milvus migrans migrans (Boddaert)	Black Kite
Accipiter nisus nisus (Linnaeus)	Sparrowhawk
Buteo lineatus (Gmelin)	Red-shouldered Hawk
Buteo buteo buteo (Linnaeus)	Buzzard
Aquila chrysaëtos chrysaëtos (Linnaeus)	Golden Eagle
Haliæetus leucogaster (Gmelin)	White-bellied Sea Eagle
Haliæetus albicilla (Linnaeus)	White-tailed Sea Eagle
Circus pygargus (Linnaeus)	Montagu's Harrier
Terathopius ecaudatus (Daudin)	Bateleur Eagle

Family FALCONIDAE

Milvago chimango (Vieillot)	Chimango
Phalcobaenus australis (Gmelin)	Forster's Milvago
Falco rusticolus candicans Gmelin	Greenland Falcon
Falco peregrinus peregrinus Tunstall	Peregrine Falcon
Falco columbarius aesalon Tunstall	Merlin
Falco naumanni naumanni Fleischer	Lesser Kestrel

Order GALLIFORMES

Family MEGAPODIIDAE

Alectura lathami J. E. Gray	Brush Turkey

Family CRACIDAE

Penelopina nigra (Fraser)	Black Penelope

Order GALLIFORMES (*continued*)

Family TETRAONIDAE

Tetrao urogallus urogallus Linnaeus	Capercaillie
Lyrurus tetrix britannicus Witherby & Lönnberg	British Black Grouse
Lagopus scoticus (Latham)	Red Grouse
Lagopus lagopus lagopus (Linnaeus)	Willow Grouse
Canachites franklinii (Douglas)	Franklin's Grouse
Bonasa umbellus umbellus (Linnaeus)	Ruffed Grouse
Pedioecetes phasianellus campestris Ridgway	Prairie Sharp-tailed Grouse
Tympanuchus cupido pinnatus (Brewster)	Greater Prairie Chicken
Tympanuchus pallidicinctus (Ridgway)	Lesser Prairie Chicken
Centrocercus urophasianus (Bonaparte)	Sage Hen

Family PHASIANIDAE

Colinus virginianus (Linnaeus)	Bob-white Quail
Odontophorus gujanensis marmoratus (Gould)	Wood Quail
Perdix perdix (Linnaeus)	Partridge
Ptilopachus petrosus (Gmelin)	Stone Pheasant
Tragopan satyra (Linnaeus)	Crimson Horned Pheasant
Lophura rufa (Raffles)	Crested Fireback Pheasant
Gallus gallus (Linnaeus)	Domestic Fowl
Lobiophasis bulweri Sharpe	Bulwer's Pheasant
Phasianus colchicus Linnaeus	Pheasant
Chrysolophus amherstiae (Leadbeater)	Amherst Pheasant
Polyplectron bicalcaratum (Linnaeus)	Grey Peacock Pheasant
Argusianus argus argus (Linnaeus)	Argus Pheasant
Argusianus argus grayi (Elliot)	Bornean Argus Pheasant
Pavo cristatus Linnaeus	Peacock

Family NUMIDIDAE

Numida meleagris (Linnaeus)	Guineafowl

Family MELEAGRIDIDAE

Meleagris gallopavo Linnaeus	Turkey
Meleagris gallopavo silvestris Vieillot	Eastern Turkey
Meleagris gallopavo gallopavo Linnaeus	Mexican Turkey

Order GRUIFORMES

Family TURNICIDAE

Turnix sylvatica lepurana (A. Smith)	Kurriehane Button Quail

Order GRUIFORMES (*continued*)

Family GRUIDAE

Grus grus grus (Linnaeus)	Common Crane
Grus canadensis canadensis (Linnaeus)	Little Brown Crane
Grus americana (Linnaeus)	Whooping Crane
Grus antigone antigone (Linnaeus)	Sarus Crane
Grus rubicunda (Perry)	Native Companion
Anthropoides virgo (Linnaeus)	Demoiselle Crane
Anthropoides paradisea (Lichtenstein)	Stanley Crane

Family PSOPHIIDAE

Psophia sp.	Trumpeter

Family RALLIDAE

Rallus limicola limicola Vieillot	Virginia Rail
Rallus aquaticus aquaticus Linnaeus	Water Rail
Aramides cajanea cajanea (P. L. S. Müller)	Wood Rail
Aramides ypecaha (Vieillot)	Ypecaha Rail
Crex crex (Linnaeus)	Corncrake
Gallinula chloropus chloropus (Linnaeus)	Moorhen or Waterhen
Fulica atra atra Linnaeus	Coot

Family EURYPYGIDAE

Eurypyga helias (Pallas)	Sun Bittern

Family OTIDAE

Otis tarda tarda Linnaeus	Great Bustard
Choriotis australis (J. E. Gray)	Australian Bustard

Order CHARADRIIFORMES

Family ROSTRATULIDAE

Rostratula benghalensis benghalensis (Linnaeus)	Painted Snipe

Family HAEMATOPODIDAE

Haematopus ostralegus occidentalis Neumann	British Oystercatcher or Seapie

Family CHARADRIIDAE

Chettusia gregaria (Pallas)	Sociable Plover
Vanellus vanellus (Linnaeus)	Lapwing
Belonopterus chilensis cayennensis (Gmelin)	Spur-winged Lapwing
Squatarola squatarola (Linnaeus)	Grey Plover
Pluvialis apricaria apricaria (Linnaeus)	Northern Golden Plover
Pluvialis apricaria oreophilos A. C. Meinertzhagen	Southern Golden Plover

Order CHARADRIIFORMES (*continued*)

Family CHARADRIIDAE (*continued*)

Charadrius hiaticula hiaticula Linnaeus	Ringed Plover
Charadrius alexandrinus alexandrinus Linnaeus	Kentish Plover
Charadrius pecuarius Temminck	Kittlitz's Sand Plover
Charadrius vociferus vociferus Linnaeus	Killdeer Plover
Eudromias morinellus (Linnaeus)	Dotterel

Family SCOLOPACIDAE

Numenius arquata arquata (Linnaeus)	Curlew
Limosa limosa limosa (Linnaeus)	Black-tailed Godwit
Tringa totanus (Linnaeus)	Redshank
Tringa nebularia (Gunnerus)	Greenshank
Tringa ocrophus Linnaeus	Green Sandpiper
Capella media (Latham)	Great Snipe
Capella gallinago gallinago (Linnaeus)	Common Snipe
Scolopax rusticola Linnaeus	Woodcock
Calidris canutus canutus (Linnaeus)	Knot
Calidris canutus rufus (Wilson)	Americàn or Red Knot
Crocethia alba (Pallas)	Sanderling
Erolia temminckii (Leisler)	Temminck's Stint
Erolia bairdii (Coues)	Baird's Sandpiper
Erolia maritima (Brünnich)	Purple Sandpiper
Erolia alpina schinzii (C. L. Brehm)	Southern Dunlin
Tryngites subruficollis (Vieillot)	Buff-breasted Sandpiper
Philomachus pugnax (Linnaeus)	Ruff

Family RECURVIROSTRIDAE

Himantopus himantopus himantopus (Linnaeus)	Black-winged Stilt
Himantopus himantopus novaezelandiae Gould	Pied Stilt
Himantopus himantopus mexicanus (P. L. S. Müller)	Black-necked Stilt
Himantopus himantopus melanurus Vieillot	Brazilian Stilt
Recurvirostra avosetta Linnaeus	Avocet
Recurvirostra americana Gmelin	American Avocet

Family PHALAROPODIDAE

Lobipes lobatus (Linnaeus)	Red-necked Phalarope

Family BURHINIDAE

Burhinus oedicnemus oedicnemus (Linnaeus)	Stone Curlew

Order CHARADRIIFORMES (*continued*)

Family GLAREOLIDAE
   Glareola pratincola pratincola (Linnaeus)   Collared Pratincole
   Glareola lactea Temminck   Little Pratincole

Family THINOCORIDAE
   Thinocorus rumicivorus patagonicus   Patagonian Seed-snipe
   Reichenow

Family STERCORARIIDAE
   Catharacta skua skua Brünnich   Great Skua
   Catharacta skua intercedens Mathews   Kerguelen Skua
   Stercorarius longicaudus Vieillot   Long-tailed Skua

Family LARIDAE
   Larus canus canus Linnaeus   Common Gull
   Larus argentatus argentatus Pontoppidan   Herring Gull
   Larus fuscus graellsii A. E. Brehm   British Lesser Black-backed
       Gull

   Larus marinus Linnaeus   Great Black-backed Gull
   Larus atricilla Linnaeus   Laughing Gull
   Larus ridibundus Linnaeus   Black-headed Gull
   Larus minutus Pallas   Little Gull
   Rissa tridactyla tridactyla (Linnaeus)   Kittiwake
   Hydroprogne tschegrava tschegrava   Caspian Tern
      (Lepechin)
   Sterna hirundo hirundo Linnaeus   Common Tern
   Sterna paradisaea Pontoppidan   Arctic Tern
   Sterna dougallii dougallii Montagu   Roseate Tern
   Sterna fuscata fuscata Linnaeus   Sooty Tern
   Sterna albifrons albifrons Pallas   Little Tern
   Thalasseus sandvicensis sandvicensis   Sandwich Tern
      (Latham)
   Anous stolidus stolidus (Linnaeus)   Noddy Tern

Family RYNCHOPIDAE
   Rynchops nigra nigra Linnaeus   Black Skimmer

Family ALCIDAE
   Pinguinus impennis (Linnaeus)   Great Auk
   Alca torda Linnaeus   Razorbill
   Uria aalge aalge (Pontoppidan)   Northern Guillemot
   Uria aalge albionis Witherby   Southern Guillemot
   Cepphus grylle grylle (Linnaeus)   Black Guillemot
   Fratercula arctica grabae (C. L. Brehm)   Southern Puffin

Order COLUMBIFORMES
Family PTEROCLIDIDAE — Sandgrouse
Family RAPHIDAE
  Raphus cucullatus (Linnaeus) — Dodo
Family COLUMBIDAE
  Columba livia Gmelin — Domestic Pigeon
  Columba livia livia Gmelin — Rock Dove
  Columba oenas oenas Linnaeus — Stock Dove
  Columba palumbus palumbus Linnaeus — Wood Pigeon
  Ectopistes migratoria (Linnaeus) — Passenger Pigeon
  Nesopelia galapagoensis (Gould) — Galapagos Dove
  Streptopelia decaocto (Frivaldszky) — Ring Dove
  Metriopelia ceciliae (Lesson) — Red-throated Ground Dove

Order PSITTACIFORMES
Family PSITTACIDAE
  Nestor notabilis Gould — Kea
  Trichoglossus chlorolepidotus (Kuhl) — Scaly-breasted Lorikeet
  Myiopsitta monachus (Boddaert) — Monk Parakeet
  Psephotus chrysopterygius Gould — Golden-shouldered Parakeet
  Melopsittacus undulatus (Shaw) — Budgerigar

Order CUCULIFORMES
Family CUCULIDAE
  Cuculus canorus canorus Linnaeus — Cuckoo
  Coccyzus cinereus Vieillot — Ash-coloured Cuckoo
  Coccyzus erythrophthalmus (Wilson) — Black-billed Cuckoo
  Coccyzus americanus (Linnaeus) — Yellow-billed Cuckoo
  Crotophaga ani Linnaeus — Smooth-billed Ani
  Crotophaga sulcirostris Swainson — Groove-billed Ani
  Guira guira (Gmelin) — White Ani
  Geococcyx californianus (Lesson) — Roadrunner
  Centropus bengalensis javanensis (Dumont) — Malay Coucal

Order STRIGIFORMES
Family TYTONIDAE
  Tyto alba alba (Scopoli) — White-breasted Barn Owl
Family STRIGIDAE
  Bubo bubo bubo (Linnaeus) — Eagle Owl
  Nyctea scandica (Linnaeus) — Snowy Owl
  Strix aluco sylvatica Shaw — British Tawny Owl

Order STRIGIFORMES (*continued*)
  Family STRIGIDAE (*continued*)
    Strix varia helveola Bangs — Texas Barred Owl
    Asio otus otus (Linnaeus) — Long-eared Owl
    Asio flammeus flammeus (Pontoppidan) — Short-eared Owl
Order CAPRIMULGIFORMES
  Family PODARGIDAE
    Podargus strigoides (Latham) — Frogmouth
  Family CAPRIMULGIDAE
    Chordeiles acutipennis texensis Lawrence — Texas Nighthawk
    Chordeiles minor (Forster) — Nighthawk
    Phalaenoptilus nuttallii californicus Ridgway — Dusky Poorwill
    Caprimulgus europaeus europaeus Linnaeus — Nightjar
Order APODIFORMES
  Family APODIDAE
    Apus apus apus (Linnaeus) — Swift
  Family TROCHILIDAE
    Phaethornis sp. — Hermit Humming-bird
    Florisuga mellivora (Linnaeus) — Great Jacobin Humming-bird
    Amazilia tzacatl tzacatl (De la Llave) — Rieffer's Humming-bird
    Hylocharis leucotis (Vieillot) — White-eared Humming-bird
    Topaza pella pella (Linnaeus) — Guiana King Humming-bird
    Selasphorus alleni Henshaw — Allen's Humming-bird
    Selasphorus platycercus platycercus (Swainson) — Broad-tailed Humming-bird
    Calypte anna (Lesson) — Anna's Humming-bird
    Loddigesia mirabilis (Bourc.) — Loddige's Spatule-tail
    Anthracothorax nigricollis nigricollis (Vieillot) — Black-throated Humming-bird
Order CORACIIFORMES
  Family ALCEDINIDAE
    Megaceryle alcyon (Linnaeus) — Belted Kingfisher
    Alcedo atthis ispida Linnaeus — European Kingfisher
  Family MOMOTIDAE
    Momotus bahamensis Swainson — Swainson's Motmot
  Family MEROPIDAE
    Merops apiaster Linnaeus — European Bee-eater

Order CORACIIFORMES (continued)
Family BUCEROTIDAE
  Buceros rhinocerus Linnaeus — Rhinoceros Hornbill
  Dichoceros bicornis (Linnaeus) — Indian Great Hornbill
  Rhytidoceros plicatus (Forster) — New Guinea Hornbill
  Lophoceros melanoleucus (Lichtenstein) — Crowned Hornbill

Order PICIFORMES
Family GALBULIDAE
  Galbula melanogenia Sclater — Black-chinned Jacamar
Family CAPITONIDAE
  Megalaima marshallorum Swinhoe — Great Himalayan Barbet
  Trachyphonus d'arnaudii böhmi Fischer & Reichenow — East African Barbet
Family INDICATORIDAE
  Indicator indicator (Gmelin) — Black-throated Honey-guide
Family PICIDAE
  Colaptes auratus (Linnaeus) — Yellow-shafted Flicker
  Colaptes auratus luteus Bangs — Northern Flicker
  Colaptes cafer collaris Vigors — Red-shafted Flicker
  Dryobates minor comminutus (Hartert) — British Lesser Spotted Woodpecker
  Campephilus principalis (Linnaeus) — Ivory-billed Woodpecker
  Jynx torquilla torquilla Linnaeus — Wryneck

Order PASSERIFORMES
Family FURNARIIDAE
  Furnarius rufus (Gmelin) — Red Ovenbird
Family COTINGIDAE
  Rupicola rupicola (Linnaeus) — Cock of the Rock
  Cotinga nattererii (Boissoneau) — Natterer's Chatterer
  Procnias averano carnobarba (Cuvier) — Black-winged Bellbird
  Procnias tricarunculatus (Verreaux) — Costa-Rican Bellbird
Family PIPRIDAE
  Pipra mentalis minor Hartert — Yellow-thighed or Red-headed Manakin
  Chiroxiphia caudata (Shaw) — Long-tailed Manakin
  Manacus manacus vitellinus (Gould) — Gould's Manakin
  Manacus manacus gutturosus (Desmarest) — Bearded Manakin

ABD

Order PASSERIFORMES (*continued*)

Family TYRANNIDAE
Knipolegus hudsoni (Sclater) — Hudson's Black Tyrant Flycatcher
Lichenops perspicillata (Gmelin) — Silverbill
Muscivora tyrannus (Linnaeus) — Scissortail

Family MENURIDAE
Menura novaehollandiae Latham — Lyrebird

Family ALAUDIDAE
Alauda arvensis arvensis Linnaeus — Skylark
Mirafra fasciolata
Lullula arborea arborea (Linnaeus) — Woodlark

Family HIRUNDINIDAE
Hirundo rustica rustica Linnaeus — Swallow
Riparia riparia riparia (Linnaeus) — Sand Martin

Family CAMPOPHAGIDAE
Pericrocotus sp. — Minivet

Family ORIOLIDAE
Oriolus oriolus oriolus (Linnaeus) — Golden Oriole

Family CORVIDAE
Corvus frugilegus frugilegus Linnaeus — Rook
Corvus corax corax Linnaeus — Raven
Corvus corone corone Linnaeus — Carrion Crow
Corvus splendens splendens Vieillot — House Crow
Corvus brachyrhynchos Brehm — American Crow
Corvus monedula spermologus Vieillot — Jackdaw
Nucifraga caryocatactes caryocatactes (Linnaeus) — Thick-billed Nutcracker
Pica pica pica (Linnaeus) — Magpie
Urocissa melanocephala occipitalis (Blyth) — Red-billed Blue Magpie
Garrulus glandarius glandarius (Linnaeus) — Continental Jay
Garrulus glandarius rufitergum Hartert — British Jay
Psilorhinus mexicanus cyanogenys Sharpe — Central American Jay
Pyrrhocorax pyrrhocorax pyrrhocorax (Linnaeus) — Chough
Corcorax melanorhamphus (Vieillot) — White-winged Chough

Family PARADISEIDAE
Craspedophora magnifica (Vieillot) — Magnificent Riflebird
Seleucides melanoleucus melanoleucus Finsch — Long-tailed Bird of Paradise
Epimachus meyeri meyeri (Daudin) — Twelve-wired Bird of Paradise

Order PASSERIFORMES (*continued*)

Family PARADISEIDAE (*continued*)

Paradisea apoda Linnaeus	Greater Bird of Paradise
Paradisea apoda salvadorii	Count Salvadori's Bird of Paradise
Paradisea raggiana Sclater	Count Raggi's Bird of Paradise
Paradisea minor Shaw	Lesser Bird of Paradise
Trichoparadisea gulielmi Cabanis	Emperor of Germany's Bird of Paradise
Paradisornis rudolphi Finsch	Prince Rudolph's Bird of Paradise
Cicinnurus regius (Linnaeus)	King Bird of Paradise
Diphyllodes magnificus (Forster)	Magnificent Bird of Paradise
Diphyllodes magnificus hunsteini A. B. Meyer	Hunstein's Bird of Paradise
Schlegelia wilsoni (Cassin)	Wilson's Bird of Paradise
Semioptera wallacei Gray	Wallace's Standardwing
Semioptera wallacei halmaherae Salvadori	Halmahera Standardwing
Parotia sefilata (Forster)	Six-wired Bird of Paradise
Lophorina superba (Forster)	Superb Bird of Paradise
Lophorina superba minor Ramsay	Lesser Superb Bird of Paradise
Ptilonorhynchus violaceus (Vieillot)	Satin Bowerbird
Sericulus chrysocephalus (Lewin)	Regent Bowerbird

Family PARIDAE

Parus major newtoni Prazak	British Great Tit
Parus caeruleus obscurus Prazak	British Blue Tit
Parus palustris dresseri Stejneger	British Marsh Tit
Psaltriparus melanotis (Hartlaub)	Black-eared Bush Tit
Psaltriparus minimus (Townsend)	Little Bush Tit
Panurus biarmicus biarmicus (Linnaeus)	Bearded Tit

Family SITTIDAE

Sitta europaea affinis Blyth	British Nuthatch

Family CERTHIIDAE

Certhia familiaris occidentalis Ridgway	Californian Tree Creeper

Family CHAMAEIDAE

Chamaea fasciata (Gambel)	Wren-tit

Family TIMALIIDAE

Psophodes olivaceus (Latham)	Whipbird

Order PASSERIFORMES (*continued*)
Family TIMALIIDAE (*continued*)
   Pomatostomus temporalis (Vigors &   Grey-crowned Babbler
    Horsfield)
   Yuhina brunneiceps Grant
   Leiothrix lutea (Scopoli)   Pekin Robin
Family PYCNONOTIDAE   Bulbuls
Family CINCLIDAE
   Cinclus cinclus hibernicus Hartert   Irish Dipper
Family TROGLODYTIDAE
   Heleodytes bicolor Pelzeln   Cactus Wren
   Heleodytes zonatus zonatus (Lesson)   Banded Cactus Wren
   Heleodytes nuchalis (Cabanis)
   Troglodytes aedon Vieillot   House Wren
   Troglodytes troglodytes troglodytes   Wren
    (Linnaeus)
Family MIMIDAE
   Mimus polyglottos (Linnaeus)   Common Mockingbird
   Nesomimus melanotis (Gould)   Galapagos Mockingbird
Family TURDIDAE
   Turdus merula merula Linnaeus   Blackbird
   Turdus torquatus torquatus Linnaeus   Ring Ousel
   Turdus ericetorum ericetorum Turton   British Song Thrush
   Hylocichla minima minima (Lafresnáye)   Bicknell's Thrush
   Hylocichla minima aliciae (Baird)   Grey-cheeked Thrush
   Monticola saxatilis (Linnaeus)   Rock Thrush
   Ephthianura albifrons (Jardine & Selby)   White-fronted Chat
   Cyanosylvia svecica svecica (Linnaeus)   Red-spotted Bluethroat
   Luscinia megarhyncha megarhyncha Brehm   Nightingale
   Erithacus rubecula melophilus Hartert   British Robin
   Copsychus saularis (Linnaeus)   Magpie Robin
   Oenanthe oenanthe oenanthe (Linnaeus)   Wheatear
Family SYLVIIDAE
   Locustella luscinoides luscinoides (Savi)   Savi's Warbler
   Locustella certhiola (Pallas)   Pallas's Grasshopper
                           Warbler
   Locustella naevia naevia (Boddaert)   Grasshopper Warbler
   Acrocephalus schoenobaenus (Linnaeus)   Sedge Warbler
   Acrocephalus palustris (Bechstein)   Marsh Warbler
   Acrocephalus scirpaceus scirpaceus   Reed Warbler
    (Hermann)

Order PASSERIFORMES *(continued)*

Family SYLVIIDAE *(continued)*

Orthotomus sutorius (Forster)	Tailorbird
Sylvia communis communis Latham	Whitethroat
Sylvia curruca curruca (Linnaeus)	Lesser Whitethroat
Sylvia atricapilla atricapilla (Linnaeus)	Blackcap
Phylloscopus trochilus trochilus (Linnaeus)	Willow Warbler
Phylloscopus collybita collybita (Vieillot)	Chiffchaff
Malurus cyaneus (Latham)	Superb Blue Wren
Malurus lamberti Vigors & Horsfield	Variegated Wren

Family REGULIDAE

Regulus regulus regulus (Linnaeus)	Goldcrest

Family PRUNELLIDAE

Prunella modularis occidentalis (Hartert)	British Hedgesparrow
Prunella collaris collaris (Scopoli)	Alpine Accentor

Family MOTACILLIDAE

Motacilla alba yarrellii Gould	Pied Wagtail
Anthus trivialis trivialis (Linnaeus)	Tree Pipit
Anthus spinoletta petrosus (Montagu)	Rock Pipit

Family BOMBYCILLIDAE

Bombycilla cedrorum Vieillot	Cedar Waxwing

Family DULIDAE

Dulus dominicus dominicus (Linnaeus)	Palm Chat

Family LANIIDAE

Gymnorhina hypoleuca (Gould)	White-backed Magpie
Lanius excubitor elegans Swainson	Pallid Shrike
Lanius ludovicianus ludovicianus Linnaeus	Loggerhead Shrike
Lanius collurio collurio Linnaeus	Red-backed Shrike
Laniarius ferrugineus sublacteus (Cassin)	Zanzibar Boubou
Laniarius erythrogaster (Cretzschmar)	Abyssinian Gonolek

Family PRIONOPIDAE

Prionops plumata plumata (Shaw)	Long-crested Helmet Shrike

Family STURNIDAE

Sturnus vulgaris vulgaris Linnaeus	Starling
Pastor roseus (Linnaeus)	Rose-coloured Starling
Heteralocha acutirostris (Gould)	Huia

Family MELITHREPTIDAE

Meliphaga melanops (Latham)	Yellow-tufted Honeyeater

Order PASSERIFORMES (*continued*)
Family NECTARINIIDAE
  Arachnothera longirostris (Latham)   Little Spiderhunter

Family ZOSTEROPIDAE
  Zosterops palpebrosa japonica Temminck   Japanese Silvereye
  & Schlegel

Family COEREBIDAE
  Cyanerpes cyaneus (Linnaeus)   Yellow-winged Blue
                                                            Sugar-bird
Family COMPSOTHLYPIDAE
  Limnothlypis swainsoni (Audubon)   Swainson's Warbler
  Dendroica castanea (Wilson)   Bay-breasted Warbler
  Dendroica discolor collinsi Bailey   Florida Prairie Warbler
  Geothlypis trichas brachidactyla (Swainson)   Northern Yellowthroat

Family PLOCEIDAE
  Passer domesticus domesticus (Linnaeus)   House Sparrow
  Euplectes hordeacea hordeacea (Linnaeus)   Crimson-crowned Bishop-
                                                            bird
  Philetairus socius (Latham)   Sociable Weaverbird
  Poephila gouldiae (Gõuld)   Gouldian Weaverfinch
  Erythrura cyanovirens cyanovirens (Peale)   Samoan Red-tailed
                                                            Weaverfinch

Family ICTERIDAE
  Zarhynchus wagleri wagleri (Gray)   Wagler's Oropendola
  Xanthornus viridis (P. L. S. Müller)   Green Oropendola
  Cassidix mexicanus mexicanus (Gmelin)   Great-tailed Grackle
  Cassidix mexicanus major (Vieillot)   Boat-tailed Grackle
  Dolichonyx oryzivorus (Linnaeus)   Bobolink
  Molothrus ater (Boddaert)   North American Cowbird
  Agelaius phoeniceus (Linnaeus)   Red-winged Blackbird
  Agelaius tricolor (Audubon)   Tricoloured Redwing
  Xanthocephalus xanthocephalus (Bonaparte)   Yellow-headed Blackbird
  Icterus pyrrhopterus (Vieillot)   Chestnut-shouldered
                                                            Hangnest
  Megaquiscalus nicaraguensis (Salvin &   Nicaraguan Grackle
  Godman)

Family THRAUPIDAE
  Tachyphonus melanoleucus (Vieillot)   Black Tanager

Order PASSERIFORMES (*continued*)

Family FRINGILLIDAE

Cactospiza pallida (Sclater & Salvin)	(A Galapagos finch)
Camarhynchus parvulus (Gould)	(A Galapagos finch)
Chloris chloris chloris (Linnaeus)	Greenfinch
Coccothraustes coccothraustes coccothraustes (Linnaeus)	Hawfinch
Volatinia jacarinia splendens (Vieillot)	Glossy Grassquit
Fringilla coelebs coelebs Linnaeus	Chaffinch
Carduelis carduelis britannica (Hartert)	British Goldfinch
Carduelis caniceps Vigors	Grey-headed Goldfinch
Spinus spinus (Linnaeus)	Siskin
Serinus canaria canaria (Linnaeus)	Canary
Sicalis arvensis (Kittlitz)	Misto Fieldfinch
Loxia curvirostra curvirostra Linnaeus	Crossbill
Pyrrhula pyrrhula nesa Mathews & Iredale	British Bullfinch
Emberiza citrinella citrinella Linnaeus	Yellow Bunting
Emberiza schoeniclus schoeniclus (Linnaeus)	Reed Bunting
Plectrophenax nivalis nivalis (Linnaeus)	Snow Bunting
Calcarius lapponicus lapponicus (Linnaeus)	Lapland Bunting
Junco hyemalis hyemalis (Linnaeus)	Slate-coloured Junco
Melospiza melodia euphonia Wetmore	Mississipi Song Sparrow
Passerina cyanea (Latham)	Indigo Bunting
Passerina ciris (Linnaeus)	Painted Bunting
Pipilo erythrophthalmus erythrophthalmus (Linnaeus)	Red-eyed Towhee

Bibliography

Adams, D. K. (1931). Restatement of the problem of learning. *Brit. J. Psychol.* **22**, 150–78.

Addicott, A. B. (1938). Behavior of the bush-tit in the breeding season. *Condor*, **40**, 49–63.

Ahlquist, H. (1937). Psychologische Beobachtungen an einigen Jungvögeln der Gattungen *Stercorarius, Larus,* und *Sterna. Acta Soe. Fauna Flora Fennica,* **60**, 162–78.

Aiken, C. E. H. and Warren, E. (1914). The birds of El Paso county, Colorado. II. *Colo. Coll. Publ. Sci.* **12**, 497–603.

Ainslie, J. A. and Atkinson, R. (1937). On the breeding habits of Leach's fork-tailed petrel. *Brit. Birds,* **30**, 234–48, 276–7.

Aldrich, E. C. (1935). Nesting of the dusky poorwill. *Condor,* **37**, 49–55.

Allee, W. C. (1931). *Animal Aggregations: A Study in General Sociology.* Chicago.

Allee, W. C. (1935). Relatively simple animal aggregations. Chap. xix in *Handbook of Social Psychology,* ed. C. Murchison. Clark Univ. Press.

Allee, W. C. (1936). Analytical studies of group behavior in birds. *Wilson Bull.* **48**, 145–51.

Allee, W. C. (1939). *The Social Life of Animals.* London and Toronto.

Allee, W. C. and Collias, N. (1938). Influence of injected male hormone on the social hierarchy in small flocks of hens. *Anat. Rec.* **72**, suppl. 61.

Allee, W. C., Collias, N. E. and Lutherman, C. Z. (1939). Modification of the social order in flocks of hens by the injection of testosterone propionate. *Physiol. Zoöl.* **12**, 412–40.

Allen, A. A. (1914). The red-winged blackbird: a study in the ecology of a cat-tail marsh. *Proc. Linn. Soc. N.Y.,* nos. 24, 25, pp. 43–128.

Allen, A. A. (1934). Sex rhythm in the ruffed grouse (*Bonasa umbellus* Linn.), and other birds. *Auk,* **51**, 180–99.

Allen, A. A. and Kellogg, P. P. (1937). Recent observations on the ivory-billed woodpecker. *Auk,* **54**, 166–84.

Allen, E., ed. (1932). *Sex and Internal Secretions.* Baltimore.

Allen, E., ed. (1939). *Sex and Internal Secretions.* London.

Allen, G. M. (1925). *Birds and their Attributes.* London.

Allen, R. P. and Mangels, F. P. (1940). Studies of the nesting behavior of the black-crowned night heron. *Proc. Linn. Soc. N.Y.,* nos. 50, 51, pp. 1–28.

Alphéraky, S. N. (1906). Note on great snipe. *Field,* p. 1075.

330   BIBLIOGRAPHY

Altum, J. B. T. (1868). *Der Vogel und sein Leben*. Münster.
Alverdes, F. (1935). The behavior of mammalian herds and packs. Chap. VI in *Handbook of Social Psychology*, ed. C. Murchison. Clark Univ. Press.
Angier, R. P. (1927). The conflict theory of emotion. *Amer. J. Psychol.* **39**, 390–401.
Aristotle. Ἰστορίαι περὶ ζῴων. Cf. Thompson, D'A. W.
Armstrong, E. A. (1929). The cuckoo. *Sci. Progr.* **24**, 81–96.
Armstrong, E. A. (1940). *Birds of the Grey Wind*. Oxford and London.
Armstrong, E. A. (in the press). The ritual of the plough. *Folklore*.
Armstrong, E. A. and Phillips, G. W. (1925). Notes on the nesting of the short-eared owl in Yorkshire. *Brit. Birds*, **18**, 226–30.
Arnold, L. W. (1930). Observations upon humming-birds. *Condor*, **32**, 302–3.
Audubon, J. J. (1840–4). *The Birds of America*. New York and Philadelphia.
Aylmer, E. A. (1932). The nest of the white-bellied sea eagle. *Hong Kong Nat.* **3**, 80–1.
Aymar, G. C. (1936). *Bird Flight*. London.

Bagg, A. C. and Eliot, S. A. (1933). Courtship of the hooded merganser (*Lophodytes cucullatus*). *Auk*, **50**, 430–1.
Bagshawe, T. W. (1938). Notes on the habits of the gentoo and ringed or Antarctic penguins. *Trans. zool. Soc. Lond.* **24**, 185–306.
Baker, C. E. (1937). Cited in F. C. R. Jourdain, So-called injury-feigning among birds. III. *Oolog. Rec.* **17**, 14–16.
Baker, J. R. (1937). Light and breeding seasons. *Nature, Lond.*, **139**, 414.
Baker, J. R. (1938). The evolution of breeding seasons. *Evolution: Essays on Aspects of Evolutionary Biology presented to Prof. E. S. Goodrich*, ed. G. R. de Beer, pp. 161–77. Oxford.
Baker, J. R. (1939). The relation between latitude and breeding seasons in birds. *Proc. zool. Soc. Lond.*, A, **108**, 557–82.
Baker, J. R. and Baker, I. (1936). The seasons in a tropical rain-forest (New Hebrides). Pt. II. Botany. *J. linn. Soc. (Zool.)*, **39**, 507.
Baker, J. R. and Baker, Z. (1936). The seasons in a tropical rain-forest (New Hebrides). Pt. III. Fruit-bats (Pteropidae). *J. linn. Soc. (Zool.)*, **40**, 123–41.
Baker, J. R. and Bird, T. F. (1936). The seasons in a tropical rain-forest (New Hebrides). Pt. IV. Insectivorous bats (Vespertilionidae and Rhinolophidae). *J. linn. Soc. (Zool.)*, **40**, 143–61.
Baker, S. (1935). *Nidification of Birds in the Indian Empire*, **4**, 326.
Barrows, W. B. (1913). Concealing action of the bittern. *Auk*, **30**, 187–90.

Bartram, J. (1754). Cited in G. Edwards, A letter to Mr Peter Collinson, F.R.S., concerning the pheasant of Pennsylvania, and the *Otis Minor*. *Philos. Trans. roy. Soc.* **48**, 499–503.

Bates, R. W., Riddle, O. and Lahr, E. L. (1937). The mechanism of the anti-gonad action of prolactin in adult pigeons. *Amer. J. Physiol.* **119**, 610–14.

Bayne, C. S. (1929). A nuptial display by wood-pigeons. *Brit. Birds*, **23**, 193–4.

Beebe, W. (1922). *A Monograph of the Pheasants*, **4**. London.

Beebe, W. (1926). *The Arcturus Adventure*. London and New York.

Beebe, W. (1938). *Zaca Venture*. London.

Belcher, C. F. (1930). *The Birds of Nyasaland*. London.

Belcher, C. F. and Smooker, G. D. (1936). Birds of the colony of Trinidad and Tobago. *Ibis* (13), **6**, 792–813.

Belt, T. (1874). *The Naturalist in Nicaragua*. London.

Bennett, M. A. (1939). The social hierarchy in ring doves. *Ecology*, **20**, 337–57.

Bennett, M. A. (1940). The social hierarchy in ring doves. II. The effect of treatment with testosterone propionate. *Ecology*, **21**, 148–65.

Benoit, J. (1934). Activation sexuelle obtenue chez le Canard par l'éclairement artificiel pendant la période de repos génital. *C.R. Acad. Sci., Paris*, **199**, 1671–73.

Benoit, J. (1935). Nouvelles expériences relatives à la stimulation par la lumière du développement testiculaire chez le Canard. *C.R. Acad. Sci., Paris*, **201**, 359–61.

Benoit, J. (1936). Facteurs externes et internes de l'activité sexuelle. I. Stimulation par la lumière de l'activité sexuelle chez le Canard et la Cane domestiques. *Bull. biol.* **70**, 487–533.

Benoit, J. (1937). Facteurs externes et internes de l'activité sexuelle. II. Étude du mécanisme de la stimulation par la lumière de l'activité sexuelle chez le Canard et la Cane domestiques. *Bull. biol.* **71**, 393–437.

Bent, A. C. (1926). Life histories of North American marsh birds. *Smithsonian Inst. Bull. U.S. nat. Mus.* **135**.

Bent, A. C. (1932). Life histories of North American gallinaceous birds. *Smithsonian Inst. Bull. U.S. nat. Mus.* **162**.

Bent, A. C. (1940). Life histories of North American cuckoos, goatsuckers, humming-birds and their allies. *Smithsonian Inst. Bull. U.S. nat. Mus.* **176**.

Berg, B. (1931). *With the Migratory Birds to Africa*, tr. F. R. Barton. London.

Berland, L. (1927). Contributions à l'étude de la biologie des arachnides. *Arch. Zool. exp. gén.* **66**, 7–29.

Beven, J. O. (1913). Notes and observations on the painted snipe (*Rostratula capensis*) in Ceylon. *Ibis* (10), **1**, 527–34.

Bierens de Haan, J. A. (1926a). Die Balz des Argusfasans. *Biol. Zbl.* **46**, 428–35.

Bierens de Haan, J. A. (1926b). Versuche über den Farbensinn und das psychische Leben von *Octopus vulgaris*. *Z. vergl. Physiol.* **4**, 766–96.

Bierens de Haan, J. A. (1929). *Animal Psychology for Biologists*. London.

Bierens de Haan, J. A. (1940). *Die tierischen Instinkte und ihr Umbau durch Erfahrung*. Leiden.

Bigg-Wither, T. P. (1878). *Pioneering in South Brazil*. London.

Bird, C. G. and Bird, E. G. (1940). Some remarks on non-breeding in the Arctic, especially in north-east Greenland. *Ibis* (14), **4**, 671–8.

Bird, G. (1935). The nesting of the hawfinch. *Brit. Birds*, **39**, 2.

Bissonnette, T. H. (1933). Light and sexual cycles in starlings and ferrets. *Quart. Rev. Biol.* **8**, 201–8.

Bissonnette, T. H. (1936a). Sexual photoperiodicity. *J. Hered.* **27**, 171–80.

Bissonnette, T. H. (1936b). Modification of mammalian sexual cycles. V. The avenue of reception of sexually stimulating light. *J. comp. Psychol.* **22**, 93–103.

Blair, R. H. and Tucker, B. W. (1941). Nest sanitation, with additions from published sources. *Brit. Birds*, **34**, 206-15, 226–35, 250–5.

Boase, H. (1924). Courting display of the fulmar. *Brit. Birds*, **18**, 45–8.

Boase, H. (1925). Courtship of the teal. *Brit. Birds*, **19**, 162–4.

Bogoras, V. (1904–09). The Chukchee. *J. N. Pacific Exp.* **7**, 268–9.

Booth, G. A. (1913). The avocet. *Wild Life*, **2**, 148–52.

Booth, E. T. (1881–7). *Rough Notes on the Birds observed during Twenty Years*. London.

Bowles, J. H. (1925). Cited in G. M. Allen, *Birds and their Attributes*. London.

Brehm, A. E. (1911–18). *Thierleben*. Leipzig and Vienna.

Bristowe, W. S. (1929). The mating habits of spiders with special reference to the problems surrounding sex dimorphism. *Proc. zool. Soc. Lond.*, pp. 309–58.

Brock, S. E. (1910). The willow-wrens of a Lothian wood. *Zoologist* (4), **14**, 401–17.

Brooks, W. S. (1915). Note on the display of the king eider. *Bull. Mus. Comp. Zool.* **59**; cited in J. C. Phillip's *Natural History of the Ducks*, **4**, 124.

Brown, W. (1940). *Psychology and Psychotherapy*, 4th ed. London.

Brückner, G. H. (1933). Untersuchungen zur Tierpsychologie insbesondere zur Auflösung der Familie. *Z. Psychol.* **128**.

Buffon, G. L. L. (1770–83). *Histoire Naturelle des Oiseaux*. Paris.

Bullough, W. S. (1941). The effect of the reduction of light in spring on the breeding behaviour of the minnow (*Phoxinus laevis* Linn.). *Proc. zool. Soc. Lond.* A, **110**, 149–57.

Bullough, W. S. (1942). Observations on the colonies of the Arctic tern (*Sterna macrura* Naumann) on the Farne Islands. *Proc. zool. Soc. Lond.*, (in the press).

Bullough, W. S. (in the press). The reproductive cycles of the British and Continental races of the starling. *Philos. Trans. roy. Soc.*

Bullough, W. S. and Carrick, R. (1939). Spring development of the gonads of the starling (*Sturnus v. vulgaris* L.). *Nature, Lond.*, **144**, 33.

Bullough, W. S. and Carrick, R. (1940). Male behaviour of the female starling. *Nature, Lond.*, **145**, 629.

Burbridge, B. (1928). *Gorilla: Tracking and Capturing the Ape-Man of Africa.* London.

Burger, J. W. (1940). Further studies on the relation of the daily exposure to light to the sexual activation of the male starling (*Sturnus vulgaris*). *J. exp. Zool.* **84**, 351–60.

Butler, A. G. (1905). On breeding *Turnix lepurana* in German bird-rooms. *Avicult. Mag.* (2), **3**, 217–22.

Buxton, A. (1932). *Sporting Interludes at Geneva.* London.

Buxton, A. (1941). The display of the blackcock. *Country Life*, **89**, no. 2295.

Cameron, E. S. (1907). The birds of Custer and Dawson Counties, Montana. *Auk*, **24**, 241–70, 389–406.

Campbell, A. G. (1941). Courtship of the lyrebird. *Emu*, **40**, 357–64.

Cannon, W. B. (1929). *Bodily Changes in Pain, Hunger, Fear and Rage*, 2nd. ed. New York and London.

Carpenter, C. R. (1933). Psycho-biological studies in Aves. I. The effect of complete and incomplete gonadectomy on the primary sexual activity of the male pigeon. II. The effect of complete and incomplete gonadectomy on secondary sexual activity, with histological studies. *J. comp. Psychol.* **16**, 25–90.

Carpenter, C. R. (1934). A field study of the behavior and social relations of howling monkeys. *Comp. Psychol. Monogr.* **10**.

Ceni, C. (1927). L'istinto materno nel maschio e le sue basi organiche. *Riv. Biol.* **9**, fasc. 3.

Chance, E. (1922). *The Cuckoo's Secret.* London.

Chance, E. (1940). *The Truth about the Cuckoo.* London.

Chapman, A. (1893). *Wild Spain.* London.

Chapman, A. (1897). *Wild Norway.* London.

Chapman, F. M. (1908a). *Camps and Cruises of an Ornithologist.* New York.

Chapman, F. M. (1908b). A contribution to the life histories of the booby (*Sula leucogaster*) and man-o'-war bird (*Fregata aquila*). *Publ. Carneg. Instn*, **103**, 139–51.

Chapman, F. M. (1928). The nesting habits of Wagler's oropendola

(*Zarhynchus wagleri*) on Barro Colorado Island. *Bull. Amer. Mus. nat. Hist.* **58**, 123–66.

Chapman, F. M. (1929). *My Tropical Air Castle.* London and New York.

Chapman, F. M. (1932). From a tropical air castle: The courtship of Gould's manakin. *Nat. Hist. N.Y.* **32**, 470–80.

Chapman, F. M. (1935). The courtship of Gould's manakin (*Manacus v. vitellinus*) on Barro Colorado Island, Canal Zone. *Bull. Amer. Mus. nat. Hist.* **68**, 471–525.

Chapman, F. M. (1938). *Life in an Air Castle.* London and New York.

Cherrie, G. K. (1916). A contribution to the ornithology of the Orinoco Region. *Sci. Bull. Brooklyn Inst.* **2**, 133–374.

Cherry-Garrard, A. (1922). *The Worst Journey in the World.* London.

Chisholm, A. H. (1934). *Bird Wonders of Australia.* Sydney.

Christoleit, E. (1926). Bemerkungen zur Biologie der Schwäne. *J. Orn.* **74**, 464–90.

Clarke, J. D. (1940). Personal communication.

Collett, R. (1876). Cited in R. B. Sharpe and H. E. Dresser, *A History of the Birds of Europe, including all the species inhabiting the Western Palaearctic Region*, **7**, 631–39. London.

Collip, J. B. (1934). Some recent advances in the physiology of the anterior pituitary. *J. Mt. Sinai Hosp.* **1**, 28.

Colman, H. R. and Boase, H. (1925). The courting display of the red-breasted merganser on salt water. *Brit. Birds*, **18**, 313–16.

Colvin, W. (1914). The lesser prairie hen. *Outing*, **63**, 608–14.

Cott, H. B. (1940). *Adaptive Coloration in Animals.* London.

Cottam, R. (1926). Cited in T. A. Coward, *The Birds of the British Isles; their Migration and Habits.* London.

Couch, J. (1847). *Illustrations of Instinct.* Edinburgh.

Courthope, W. J. (1870). *The Paradise of Birds.* Edinburgh.

Coward, T. A. (1920). *The Birds of the British Isles and their Eggs.* London.

Coward, T. A. (1923). *Birds and their Young.* London.

Coward, T. A. (1926). *The Birds of the British Isles: their Migration and Habits*, 3rd ser. London.

Craig, W. (1909). The expression of emotion in the pigeons. I. The blond ring-dove (*Turtur risorius*). *J. comp. Neurol.* **19**, 29–80.

Craig, W. (1913). The stimulation and the inhibition of ovulation in birds and mammals. *J. Anim. Behav.* **3**, 215–21.

Crandall, L. S. (1921). The display of Prince Rudolph's bird of paradise. *Bull. N.Y. zool. Soc.* **24**, 111–13.

Crandall, L. S. (1932). Notes on certain birds of paradise. *Zoologica, N.Y.*, **11**, 77–87.

Crandall, L. S. (1937). Further notes on certain birds of paradise. *Zoologica, N.Y.*, **22**, 194.

Crandall, L. S. and Leister, C. W. (1937). Display of the magnificent rifle bird. *Zoologica, N.Y.*, **22**, 311–14.

Crawley, A. E. (1902). *The Mystic Rose: A Study of Primitive Marriage and of Primitive Thought in its Bearing on Marriage.* London.

Crawley, A. E. (1918). Processions and dances. *Encyc. Rel. Eth.* **10**, 358.

Crossland, C. (1911). Warning coloration in a nudibranch mollusc and in a chameleon. *Proc. zool. Soc. Lond.* pp. 1062–7.

Crouch, J. E. (1936). The nesting habits of the cedar waxwing. *Auk*, **53**, 1–8.

Cunningham, M. (1913). The capercaillie at home. *Avicult. Mag.* (3), **4**, 236–7.

Dante Alighieri. *Inferno*, **32**, 36.

Darling, F. F. (1937). *A Herd of Red Deer.* Oxford.

Darling, F. F. (1938). *Bird Flocks and the Breeding Cycle: A Contribution to the Study of Avian Sociality.* Cambridge.

Darling, F. F. (1939). *A Naturalist on Rona: Essays of a Biologist in Isolation.* Oxford.

Darling, F. F. (1940). *Island Years.* London.

Darwin, C. (1845). *Journal of Researches into the Natural History and Geology of the countries visited during the Voyage of H.M.S. Beagle round the World,* etc., 2nd ed, with corrections and additions. London.

Darwin, C. (1859). *The Origin of Species by Means of Natural Selection or the Preservation of Favoured Races in the Struggle of Life.* London.

Darwin, C. (1871). *The Descent of Man, and Selection in Relation to Sex.* London.

Darwin, C. (1872). *The Expression of the Emotions in Man and Animals.* 2nd ed. 1874. London.

Davis, D. E. (1940a). Social habits of the smooth-billed ani. *Auk*, **57**, 179–217.

Davis, D. E. (1940b). Social nesting habits of *Guira guira. Auk*, **57**, 472–83.

Davison, W. L. (1896). Cited in W. R. Ogilvie-Grant, *Game-Birds*, **2**, 71–3.

Dawson, W. L. (1923). *The Birds of California.* Book-lovers' ed. San Diego, Los Angeles, San Francisco.

Deane, C. D. (1939). Rook behaviour. *Irish Nat. J.* **7**, 129–37.

Delacour, J. (1938). La systématique des anatidés et leurs mœurs. *Proc. 8th int. Orn. Congr. Oxford, 1934*, pp. 225–42. Oxford.

Dembo, T. (1931). Der Ärger als dynamisches Problem. *Psychol. Forsch.* **15**, 1–144.

Dewar, D. (1908). *Birds of the Plains.* London and New York.

Dewar, D. (1913). *Glimpses of Indian Birds.* London.

Dewar, D. (1928). *Birds at the Nest.* London.

Dewar, J. M. (1920*a*). The oystercatcher's progress towards maturity. *Brit. Birds*, **13**, 207–13.

Dewar, J. M. (1920*b*). The law of territory. *Brit. Birds*, **14**, 89–90.

Dircksen, R. (1932). Die Biologie des Austernfischers, der Brandseeschwalbe und der Küstenseeschwalbe nach Beobachtungen und Untersuchungen auf Norderoog. *J. Orn.* **80**, 427–521.

Dircksen, R. (1938). *Die Insel der Vögel.* Essen.

Dodsworth, P. T. L. (1910). Mental powers of animals. *Zoologist* (4), **14**, 361–76.

Domm, L. V. (1939). Modifications in sex and secondary sexual characters in birds. Chap. v in E. Allen, *Sex and Internal Secretions*, pp. 227–327.

Domm, L. V. and van Dyke, H. B. (1932). Precocious development of sexual characters in the fowl by daily injections of hebin. I. The male. *Proc. Soc. exp. Biol., N.Y.*, **30**, 349–51.

Dubois, A. D. (1924). The nuptial song-flight of the short-eared owl (*Asio flammeus*). *Auk*, **41**, 260.

Dunn, H. T. (1904). *Recollections of Dante Gabriel Rossetti and his Circle.* London.

E. P. (1941). A nest in 6695 bits. *Countryman*, **23**, 34–55.

Elliot, D. G. (1897). *The Gallinaceous Game Birds of North America.* New York.

Ellis, H. (1922). *Studies in the Psychology of Sex*, **3**, 34.

Ellison, C. S. S. (1931). Courting display of song thrush. *Brit. Birds*, **25**, 55.

Elton, C. (1932). Territory among wood ants at Picket Hill. *J. Anim. Ecol.* **1**, 69–76.

Erickson, M. M. (1938). Territory, annual cycle, and numbers in a population of wren-tits (*Chamaea fasciata*). *Univ. Calif. Publ. Zool.* **42**, 247–334.

Evans, L. T. (1936). A study of a social hierarchy in the lizard, *Anolis carolinensis. J. genet. Psychol.* **48**, 88–111.

Evans, L. T. (1936*a*). Territorial behavior of normal and castrated females of *Anolis carolinensis. J. genet. Psychol.* **49**, 49–60.

Evans, L. T. (1937). Differential effects of the ovarian hormones on the territorial reaction time of female *Anolis carolinensis. Physiol. Zoöl.* **10**, 456–63.

Evans, L. T. (1938). Cuban field studies on territoriality of the lizard, *Anolis sagrei. J. comp. Psychol.* **25**, 97–125.

Fabre, H. (1912). *Social Life in the Insect World*, tr. B. Miall. London.

Falkenstein, J. (1879). *Die Loango-Expedition ausgesandt von der deutschen Gesellschaft zur Erforschung Aequatorial-Afrikas*, **2**. Leipzig.

Farley, F. L. (1932). Cited in A. C. Bent, Life histories of North American gallinaceous birds. *Bull. U.S. nat. Mus.* **162**, 302.

Farren, W. (1914). A heronry in southern Spain. *Wild Life*, **4**, 200–15.

Finley, W. L. (1932). Cited in A. C. Bent, Life histories of North American gallinaceous birds. *Bull. U.S. nat. Mus.* **162**, 302.

Finn, F. (1919). *Bird Behaviour.* London.

Fischel, W. (1927). Beiträge zur Soziologie des Haushuhns. *Biol. Zbl.* **47**, 678–95.

Fischel, W. (1937). L'émotion et le souvenir chez les animaux. *J. Psychol. norm. path.* **34**, 374–96.

Fisher, J. (1939). *Birds as Animals.* London.

Fisher, J. (1940*a*). The world distribution and numbers of breeding gannets. *Bull. Brit. orn. Cl.* **60**, 39–41.

Fisher, J. (1940*b*). *Watching Birds.* Harmondsworth.

Fisher, J. and·Shaw, M. (1938). Courtship display. *Zoo,* **2**, 44.

Fisher, J. and Waterston, G. (1941). The breeding distribution of the fulmar in the British Isles. *J. Anim. Ecol.* **10**, 204–72.

Forde, D. (1927). *Ancient Mariners.* London.

Françon, J. (1939). *The Mind of the Bees.* Tr. H. Eltringham. London.

Freitag, F. (1936). *Vogelring,* **8**, 8–15.

Friedmann, H. (1927). Notes on some Argentine birds. *Bull. Mus. comp. Zool.* **68**, 177.

Friedmann, H. (1930). The sociable weaverbird of South Africa. *Nat. Hist. N.Y.*, **30**, 205–12.

Friedmann, H. (1934). The instinctive emotional life of birds. *Psychoanal. Rev.* **21**, nos. 3 and 4.

Friedmann, H. (1934 *a*). The display of Wallace's standard-wing bird of paradise in captivity. *Sci. Mon., N.Y.*, **39**, 52–55.

Friedmann, H. (1935). Bird societies. Chap. v in *Handbook of Social Psychology,* ed. C. Murchison, Clark Univ. Press.

Frost, W. (1910). The cock of the rock. *Avicult. Mag.* (3), **1**, 319–24.

Fuertes, L. A. (1915). Impressions of the voices of tropical birds. *Ann. Rep. Smithsonian Inst.*, pp. 299–323.

Gadamer, H. (1858). Das Balzen der *Scolopax major. J. Orn.* **6**, 235–7.

Gannon, R. A. (1930). Observations on the satin bower-bird with regard to the material used by it in painting its bower. *Emu,* **30**, 39–41.

Garner, R. L. (1896). *Gorillas and Chimpanzees.* London.

Gesner, C. (1555). *Medici Tigurini, Historia Animalium,* **3**. Zurich.

Gifford, E. W. (1941). Taxonomy and habits of pigeons. *Auk,* **58**, 239–45.

Gillespie, T. H. (1932). *A Book of King Penguins.* London.

Goethe, F. (1937). Beobachtungen und Untersuchungen zur Biologie der Silbermöwe (*Larus a. argentatus* Pontopp.) auf der Vogelinsel Memmertsand. *J. Orn.* **85**, 1–119.

Goodfellow, W. (1910). Notes on birds of paradise. *Avicult. Mag.* (3), **1**, 277–86.

Goodfellow, W. (1915). Cited in W. R. Ogilvie-Grant, Report on the Birds collected by the British Ornithological Union Expedition and the Wollaston Expedition in Dutch New Guinea. *Ibis* (10), Jubilee suppl. no. 2, p. 21.

Gordon, S. (1907). *Birds of the Loch and Mountain*. London.

Gordon, S. (1915). *Hill Birds of Scotland*. London.

Gordon, S. (1920). *The Land of the Hills and the Glens*. London.

Gordon, S. (1921). *Wanderings of a Naturalist*. London.

Gordon, S. (1936). *Thirty Years of Nature Photography*. London.

Gould, J. (1848). *An Introduction to the Birds of Australia*. London.

Granet, M. (1926). *Danses et légendes de la Chine ancienne*. Paris.

Grey, E. (1927). *The Charm of Birds*. London.

Grimes, S. A. (1936). Injury-feigning in birds. *Auk*, **53**, 37.

Groebbels, G. (1932). *Der Vogel. Bau, Funktion, Lebenserscheinung, Einpassung.* I. Atmungswelt und Nahrungswelt; II (1937). Geschlecht und Fortpflanzung. Berlin.

Groos, K. (1898). *The Play of Animals*. New York.

Gross, A. O. (1928). The heath hen. *Mem. Boston Soc. nat. Hist.* **6**, no. 4, pp. 487–588.

Gross, A. O. (1931). *Heath Hen Report*.

Gruhl, K. (1924). Paarungs-Gewohnheit der Dipteren. *Z. wiss. Zool.* **122**, 205.

Guise, R. E. (1899). On the tribes inhabiting the mouth of the Wanigela river, New Guinea. *J. anthrop. Inst.* **28**, 209.

Gurney, J. H. (1913). *The Gannet; A Bird with a History*. London.

Guthrie-Smith, H. (1929). *Bird Life on Island and Shore*. Edinburgh.

Hadden, F. C. (1941). Midway Islands. *Hawaiian Planters' Rec.* **45**, 179–222.

Hainard, R. and Meylan, O. (1935). Notes sur le Grand Tétras. *Alauda*, **7**, 282–327.

Hall, R. (1909). Notes on the magpie (*Gymnorhina leuconota* Gld.). *Emu*, **9**, 16–21.

Hamilton, J. B. and Golden, W. R. C. (1939). Responses of the female to male hormone substances. *Endocrinology*, **25**, 737–48.

Hampe, H. (1937). Zur Biologie des Rosellasittichs, *Platycercus eximius*. *J. Orn.* **85**, 175–86.

Hann, H. W. (1939). The relation of castration to migration in birds. *Bird-Banding*, **10**, 122–4.

Harper, E. H. (1904). The fertilisation and early development of the pigeon's egg. *Amer. J. Anat.* **3**, 349–86.

Harris, H. A. (1937). Cited in F. B. Kirkman, *Bird Behaviour*, p. 106.

Hartley, P. H. T. (1933). Field notes on the little grebe. *Brit. Birds*, **27**, 82–6.

Hartley, P. H. T. (1937). Sexual habits of the little grebe. *Brit. Birds*, **30**, 266.

Hatch, P. L. (1892). *Notes on the Birds of Minnesota*. Cited in A. C. Bent, *Life Histories of North American Diving Birds*, p. 56.

Haviland, M. D. (1915). Notes on the courtship of the lapwing. *Zoologist* (4), **19**, 217–25.

Haviland, M. D. (1917). Notes on the breeding habits of the dotterel on the Yenesei. *Brit. Birds*, **11**, 6–11.

Haviland, M. D. (1926). *Forest, Steppe and Tundra: Studies in Animal Environment*. Cambridge.

Heilborn, A. (1930). *Liebesspiele der Tiere*. Berlin-Charlottenburg.

Heinroth, O. (1909). Beobachtungen bei der Zucht des Ziegenmelkers (*Caprimulgus europaeus* L.). *J. Orn.* **57**, 56–83.

Heinroth, O. (1911). Beiträge zur Biologie, namentlich Ethologie und Psychologie der Anatiden. *Verh. V. Int. orn. Kongr. Berlin*, 1910, pp. 589–702. Berlin.

Heinroth, O. (1924). Lautäusserungen der Vögel. *J. Orn.* **72**, 223–44.

Heinroth, O. (1930). Über bestimmte Bewegungsweisen bei Wirbeltieren. *S.B. Ges. naturf. Fr. Berl.* Feb.

Heinroth, O. (1938). *Aus dem Leben der Vögel*. Leipzig.

Heinroth, O. (1938a). Die Balz des Bulwerfasans, *Lobiophasis bulweri* Sharpe. *J. Orn.* **86**, 1–4.

Heinroth, O. and Heinroth, M. (1924–33). *Die Vögel Mitteleuropas in allen Lebens- und Entwicklungsstufen photographisch aufgenommen und in ihrem Seelenleben bei der Aufzucht vom Ei ab beobachtet*. Berlin.

Herrick, F. H. (1901). *The Home Life of Wild Birds*. New York.

Hertz, M. (1928). Wahrnehmungspsychologische Untersuchungen am Eichelhäher. I and II. *Z. vergl. Physiol.* **7**, 144–94, 617–56.

Hertz, M. (1929). Das optische Gestaltproblem und der Tierversuch. *Verh. dtsch. zool. Ges.* pp. 23–49.

Hill, M. and Parkes, A. S. (1934). Effect of the absence of light on the breeding season of the ferret. *Proc. roy. Soc. B*, **115**, 14–17.

Hingston, R. W. G. (1933). *The Meaning of Animal Colour and Adornment*. London.

Hollom, P. A. D. (1937). Observations on the courtship and mating of the smew. *Brit. Birds*, **31**, 106.

Holzapfel, M. (1939). Analyse des Sperrens und Pickens in der Entwicklung des Stars. *J. Orn.* **87**, 525–53.

Hoogerwerf, A. (1937). Uit het leven der witte ibissen, *Threskiornis aethiopicus melanocephalus. Limosa,* **10**, 137–46.

Hoover, E. E. and Hubbard, H. E. (1937). Modification of the sexual cycle in trout by control of light. *Copeia,* no. 4, pp. 206–10.

Hosking, E. (1941). Some notes on the long-eared owl. *Brit. Birds,* **35**, 2–8.

Hosking, E. and Newberry, C. (1940). *Intimate Sketches from Bird Life.* London.

Howard, H. E. (1907–14). *The British Warblers: a History, with Problems of their Lives.* London.

Howard, H. E. (1920). *Territory in Bird Life.* London.

Howard, H. E. (1929). *An Introduction to the Study of Bird Behaviour.* Cambridge.

Howard, H. E. (1935). *The Nature of a Bird's World.* Cambridge.

Howard, H. E. (1940). *A Waterhen's Worlds.* Cambridge.

Hudson, W. H. (1892). *The Naturalist in La Plata.* London.

Hudson, W. H. (1913). *Adventures among Birds.* London.

Hudson, W. H. (1920). *The Birds of La Plata.* London.

Hudson, W. H. (1922). *A Hind in Richmond Park.* London.

Hume, A. O. (1890). *Nests and Eggs of Indian Birds.* London.

Huxley, J. S. (1914). The courtship-habits of the great crested grebe (*Podiceps cristatus*); with an addition to the theory of sexual selection. *Proc. zool. Soc. Lond.* pp. 491–562.

Huxley, J. S. (1919). Some points in the sexual habits of the little grebe with a note on the occurrence of vocal duets in birds. *Brit. Birds,* **13**, 155–8.

Huxley, J. S. (1923). Courtship activities in the red-throated diver (*Colymbus stellatus* Pontopp.); together with a discussion on the evolution of courtship in birds. *J. linn. Soc. (Zool.),* **35**, 253–92.

Huxley, J. S. (1924). Some points in the breeding behaviour of the common heron. *Brit. Birds,* **18**, 155–63.

Huxley, J. S. (1925). The absence of 'courtship' in the avocet. *Brit. Birds,* **19**, 88–9.

Huxley, J. S. (1934). A natural experiment on the territorial instinct. *Brit. Birds,* **27**, 270–7.

Huxley, J. S. (1938a). Threat and warning coloration in birds, with a general discussion of the biological functions of colour. *Proc. 8th Int. Orn. Congr. Oxford,* 1934, pp. 430–55. Oxford.

Huxley, J. S. (1938b). The present standing of the theory of natural selection. *Evolution: Essays on Aspects of Evolutionary Biology presented to Prof. E. S. Goodrich,* ed. G. R. de Beer, pp. 11–42. Oxford.

Huxley, J. S. (1938c). Darwin's theory of sexual selection and the data subsumed by it, in the light of recent research. *Amer. Nat.* **72**, 416–33.

Huxley, J. S. and Montague, F. A. (1925). Studies on the courtship and sexual life of birds. V. The oystercatcher (*Haematopus ostralegus* L.). *Ibis* (12), **1**, 868–97.

Huxley, J. S. and Montague, F. A. (1926). Studies on the courtship and sexual life of birds. VI. The black-tailed godwit, *Limosa limosa* (L.). *Ibis* (12), **2**, 1–25.

Ingram, W. J. (1907). On the display of the king bird of paradise (*Cicinnurus regius*). *Ibis* (9), **1**, 226–9.

Ingram, G. C. S. and Salmon, H. M. (1934). *Birds in Britain Today*. London.

Jackson, F. J. (1938). *The Birds of Kenya Colony and the Uganda Protectorate*, ed. W. L. Sclater. London.

James, E. (1823). *Expedition to the Rocky Mountains, 1819–20, under Major Long*, **1**, 337.

Jensen, I. (1909). Courting and mating of *Oecanthus fasciatus* Harris. *Canad. Ent.* **41**, 25.

Johnson, G. E. and Gann, E. L. (1933). Light in relation to the sexual cycle and hibernation in the thirteen-lined ground squirrel. *Anat. Rec.* **57**, suppl. 28.

Johnson, R. A. (1941). Nesting behavior of the Atlantic murre. *Auk*, **58**, 153–63.

Jourdain, F. C. R. (1936–7). The so-called 'injury-feigning' in birds. *Oolog. Rec.* **16**, 25–37; **17**, 14–16, 71–2.

Kalmus, H. (1941). Defence of source of food by bees. *Nature, Lond.*, **148**, 228.

Karsten, R. (1935). Head-hunters of western Amazonas. *Soc. Sci. Fennica, Comm. Hum. litt.* **9**. Helsingfors.

Katz, D. (1922). Tierpsychologie und Soziologie des Menschen. *Z. Psychol.* **88**, 253–64.

Katz, D. (1926). Sozialpsychologie der Vögel. *Ergebn. Biol.* **1**, 447–78.

Katz, D. (1937). *Animals and Men*. London.

Kearton, R. (1903). *Wild Nature's Ways*. London.

Kearton, R. (1915). *Wonders of Wild Nature*. London.

Keith, D. B. (1937). The red-throated diver in North-east Land. *Brit. Birds*, **31**, 66–81.

Keith, D. B. (1938). Observations on the purple sandpiper in North-east Land. *Proc. zool. Soc. Lond.* A, **108**, 185–94.

Kendeigh, S. C. and Baldwin, S. P. (1928). Development of temperature control in nestling house wrens. *Amer. Nat.* **42**, 249–78.

Kennedy, P. G. (1935). Lark singing when attacked. *Irish Nat. J.* **5**, 262.

Kidd, B. (1921). *A Philosopher with Nature.* London.

Kirkman, F. B. (1911–13). *The British Bird Book.* London.

Kirkman, F. B. (1937). *Bird Behaviour: A Contribution based chiefly on a Study of the Black-headed Gull.* London and Edinburgh.

Kirschbaum, A. (1933). Experimental modification of the seasonal cycle of the English sparrow (*Passer domesticus*). *Anat. Rec.* **57**, suppl. 62.

Kluijver, H. N. (1933). Bijdrage tot de biologie en de ecologie van den spreeuw (*Sturnus v. vulgaris* L.) gedurende zijn voortplantingstijd. *Versl. en Meded. Plantenziektenk. Dienst, Wageningen,* **69.**

Kluijver, A. N. (1935). Waarnemingen over de levenswijze van de Spreeuw (*Sturnus v. vulgaris* L.) met behulp van geringde individuen. *Ardea,* **24**, 133–66.

Koch-Grünberg, T. (1909–10). *Zwei Jahre unter den Indianern.* Berlin.

Koehler, O. and Zagarus, A. (1937). Beiträge zum Brutverhalten des Halsbandregenpfeifers (*Charadrius h. hiaticula* L.). *Beitr. FortPflBiol. Vögel,* **13**, 1–9.

Koffka, K. (1935). *Principles of Gestalt Psychology.* London and New York.

Köhler, W. (1921). Zur Psychologie des Schimpansen. *Psychol. Forsch.* **1**, 2–46.

Köhler, W. (1927). *The Mentality of Apes,* 2nd ed. London.

Kortlandt, A. (1938). De uitdrukkingsbewegingen en -geluiden van *Phalacrocorax carbo sinensis* (Shaw and Nodder). *Ardea,* **27**, 1–40.

Kretschmer, E. (1926). Hysteria. *Nerv. and Ment. Disorder Monogr.* (1), **44.** Washington.

Kropotkin, P. (1904). *Mutual Aid: A Factor of Evolution,* revised ed. London.

Küchler, W. (1935). Jahreszyklische Veränderungen im histologischen Bau der Vogelschilddrüse. *J. Orn.* **83**, 414–61.

Lack, D. (1932). Some breeding habits of the European nightjar. *Ibis* (13), **2**, 266–84.

Lack, D. (1933). Nesting conditions as a factor controlling breeding time in birds. *Proc. zool. Soc. Lond.* pp. 231–7.

Lack, D. (1935). Territory and polygamy in a bishop-bird, *Euplectes hordeacea hordeacea* (Linn.). *Ibis* (13), **5**, 817–36.

Lack, D. (1938). Display of green sandpiper. *Brit. Birds,* **32**, 86.

Lack, D. (1939*a*). The behaviour of the robin. Pt. I. The life history with special reference to aggressive behaviour, sexual behaviour and territory. Pt. II. A partial analysis of aggressive and recognitional behaviour. *Proc. zool. Soc. Lond.* A, **109**, 169–78.

Lack, D. (1939 b). The display of the blackcock. *Brit. Birds*, **32**, 290–303.

Lack, D. (1940 a). Courtship feeding in birds. *Auk*, **57**, 169–78.

Lack, D. (1940 b). Evolution of the Galapagos finches. *Nature, Lond.*, **146**, 324–5.

Lack, D. (1940 c). The releaser concept in bird behaviour. *Nature, Lond.*, **145**, 107.

Lack, D. (1940 d). Observations on captive robins. *Brit. Birds*, **33**, 262–70.

Lack, D. (1940 e). Pair-formation in birds. *Condor*, **42**, 269–86.

Lack, D. (1941). Notes on territory, fighting and display in the chaffinch. *Brit. Birds*, **34**, 216–19.

Lack, D. (1941 a). The Galapagos finches (Geospizinae): a study in variation. *Proc. Calif. Acad. Sci.* (in the press).

Lack, D. (1941 b). Some aspects of instinctive behaviour and display in birds. *Ibis* (14), **5**, 407–41.

Lack, D. L. and Emlen, J. T. (1939). Observations on breeding behavior in tricolored redwings. *Condor*, **41**, 225–30.

Lack, D. L. and Lack, L. (1933). Territory reviewed. *Brit. Birds*, **27**, 174–99.

Lagrange, F. (1889). *The Physiology of Bodily Exercise*, Eng. tr. London.

Leonard, S. (1939). Induction of singing in female canaries by injections of male hormone. *Proc. Soc. exp. Biol.*, *N.Y.*, **41**, 229–30.

Levick, G. M. (1914). *Antarctic Penguins: A Study of their Social Habits.* London.

Linsdale, J. M. (1938). Environmental responses of vertebrates in the Great Basin. *Amer. Midl. Nat.* **19**, 1–206.

Lloyd, B. (1933). The courtship and display of wheatears. *Trans. Herts. Nat. Hist. Soc.* **19**, 135–9.

Lloyd, L. (1867). *Game Birds and Wildfowl of Sweden and Norway.* London.

Lloyd, L. (1885). *The Field Sports of the North of Europe.* London.

Locket, G. H. (1923). Mating habits of Lycosidae. *Nat. Hist.*, *N.Y.*, **9**, 493–502.

Lockley, R. M. (1934). On the breeding habits of the puffin: with special reference to its breeding and fledging periods. *Brit. Birds*, **27**, 214–23.

Lockley, R. M. (1939). The sea-bird as an individual: Results of ringing experiments. *Smithsonian Inst. Report*, pp. 341–53. Reprinted from *Proc. roy. Inst.* **30**, pt. 3.

Lockley, R. M. (1941). *Shearwaters.* London.

Lorenz, K. (1931). Beiträge sur Ethologie sozialer Corviden. *J. Orn.* **79**, 67–127.

Lorenz, K. (1935). Der Kumpan in der Umwelt des Vogels: Der Artgenosse als auslösendes Moment sozialer Verhaltungsweisen. *J. Orn.* **83**, 137–213, 289–413.

Lorenz, K. (1937). The companion in the bird's world. *Auk*, **54**, 245–73.

Lorenz, K. (1937*a*). Über die Bildung des Instinktbegriffes. *Naturwissenschaften*, **25**, 289–300, 307–18, 324–31.

Lorenz, K. (1937*b*). Über den Begriff der Instinkthandlung. *Folia Biotheoretica*, **2**, 17–50.

Lorenz, K. (1938). A contribution to the comparative sociology of colonialnesting birds. *Proc. 8th Int. orn. Congr. Oxford*, 1934, pp. 207–18. Oxford.

Lorenz, K. (1939). Vergleichende Verhaltensforschung. *Zool. Anz. Suppl.* (*Verh. dtsch. zool. Ges.* 41), **12**, 69–102.

Lowe, F. A. (1934). *Days with Rarer Birds.* London.

Lumholtz, C. (1903). *Unknown Mexico*, **1**, 330.

Lynes, H. (1910). Manœuvres of lapwing in defence of young. *Brit. Birds*, **4**, 157.

Lynes, H. (1913). Early 'drumming' of the snipe and its significance. *Brit. Birds*, **6**, 354–9.

Lynes, H. (1938). A contribution to the ornithology of the southern Congo basin. *Rev. Zool. Bot. afr.* **31**, 1–128.

McCabe, T. T. (1932). Cited in A. C. Bent, Life histories of North American gallinaceous birds. *Bull. U.S. nat. Mus.* **162**, 136.

MacCurdy, J. T. (1925). *The Psychology of Emotion, Morbid and Normal.* London and New York.

McIlhenny, E. A. (1937). Life history of the boat-tailed grackle in Louisiana. *Auk*, **54**, 274–95.

McIlhenny, E. A. (1940). The sex ratio in wild birds. *Auk*, **57**, 85–93.

Maclaren, J. (1926). *My Crowded Solitude.* New York.

Makkink, G. F. (1936). An attempt at an ethogram of the European avocet (*Recurvirostra avosetta* L.), with ethological and psychological remarks. *Ardea*, **25**, 1–63.

Manniche, A. L. V. (1910). Terrestrial mammals and birds of north-east Greenland. Biological observations. *Medd. Grønland*, **45**, 1–199.

Marais, E. (1937). *The Soul of the White Ant.* London.

Marples, G. and Marples, A. (1934). *Sea Terns or Sea Swallows.* London.

Marshall, A. J. (1934). Notes on the satin bower bird in south-eastern Queensland. *Emu*, **34**, 57–61.

Marshall, F. H. A. (1922). *The Physiology of Reproduction.* London.

Marshall, F. H. A. (1936). Sexual periodicity and the causes which determine it. Croonian Lecture. *Philos. Trans. roy. Soc.* B, **226**, 423–56.

Marshall, F. H. A. (1942). Exteroceptive factors in sexual periodicity. *Biol. Rev.* **17**, 68–90.

Marshall, F. H. A. and Bowden, F. P. (1934). The effect of irradiation with different wave lengths on the oestrous cycle of the ferret, with remarks on the factors controlling sexual periodicity. *J. exp. Biol.* **11**, 409–22.

Marshall, F. H. A. and Bowden, F. P. (1936). The further effects of irradiation on the oestrous cycle of the ferret. *J. exp. Biol.* **13**, 383–6.

Marshall, F. H. A. and Walton, A. (1936). Cited as Unpublished in Marshall, 1936, *q.v.*

Martin, A. (1890). *Home Life on an Ostrich Farm.* London.

Martin, M. (1703). *Description of the Western Islands of Scotland.* In Pinkerton, *Collection of Voyages,* 1809.

Maslow, A. H. (1934). Dominance and social behavior in monkeys. *Psychol. Bull.* **31**, 688.

Maslow, A. H. (1935). Individual psychology and the social behavior of monkeys and apes. *Int. J. Indiv. Psychol.* **1**, 47–59.

Maslow, A. H. (1936). The rôle of dominance in the social and sexual behavior of infra-human primates. III. A theory of sexual behavior of infra-human primates. *J. genet. Psychol.* **48**, 310–38.

Massingham, H. J. (1931). *Birds of the Sea-shore.* London.

Masure, R. H. and Allee, W. C. (1934a). The social order in flocks of the common chicken and the pigeon. *Auk,* **51**, 306–27.

Masure, R. H. and Allee, W. C. (1934b). Flock organisation of the shell parakeet *Melopsittacus undulatus* Shaw. *Ecology,* **15**, 388–98.

Matthews, L. H. (1929). The birds of South Georgia. *Discovery Reports,* **1**.

Matthews, L. H. (1939). Visual stimulation and ovulation in pigeons. *Proc. roy. Soc.* B, **126**, 557–60.

Mayr, E. (1935). Bernard Altum and the territory theory. *Proc. Linn. Soc. N.Y.,* nos. 45, 46.

Mayr, E. (1939). The sex ratio in wild birds. *Amer. Nat.* **73**, 156–79.

Mendall, H. L. (1937). Nesting of the bay-breasted warbler. *Auk,* **54**, 429–39.

Michels, R. (1914). *Sexual Ethics,* Eng. tr. of *I limiti della morale sessuale.* Turin.

Michener, H. and Michener, J. R. (1935). Mockingbirds, their territories and individualities. *Condor,* **37**, 97–140.

Millais, J. G. (1902). *The Natural History of the British Surface-feeding Ducks.* London.

Millais, J. G. (1909). *The Natural History of British Game Birds.* London.

Millais, J. G. (1913). *British Diving Ducks.* London.

Miller, A. H. (1931). Systematic revision and natural history of the American shrikes (*Lanius*). *Univ. Calif. Publ. Zool.* **38**, 11–242.

Miller, A. H. (1937). A comparison of the behavior of certain North American and European shrikes. *Condor,* **39**, 119–22.

Miller, W. B. (1936). Flying the Pacific. *Nat. geogr. Mag.* **70**, 689–707.

346 BIBLIOGRAPHY

Miyazaki, H. (1934). On the relation of the daily period to the sexual maturity and moulting of *Zosterops palpebrosa japonica*. *Sci. Rep. Tôhoku Univ.*, Biol. ser., **9**, 183–203.

Moffat, C. B. (1924). Notes on some ruffs in the Zoo. *Irish Nat.* **33**, 25–9.

Moffat, C. B. (1940). The notes of the barn owl. *Irish Nat. J.* **7**, 289–92.

Montagu, G. (1802) (Suppl. 1813). *Ornithological Dictionary, or Alphabetical Synopsis of British Birds.* London.

Moreau, R. E. (1936). Breeding seasons of birds in East African evergreen forest. *Proc. zool. Soc. Lond.*, pp. 631–53.

Moreau, R. E. (1941). Duetting in birds. *Ibis* (14), **5**, 176–77.

Moreau, R. E. and Moreau, W. M. (1937). Biological and other notes on East African birds. *Ibis* (13), **1**, 170.

Moreau, R. E. and Moreau, W. M. (1938). Comparative breeding ecology of two species of *Euplectes* (Bishop-birds) in Usambara. *J. Anim. Ecol.* **7**, 314–27.

Moreau, R. E. and Moreau, W. M. (1939). Observations on some East African birds. *Ibis* (14), **3**, 296–323.

Moreau, R. E. and Moreau, W. M. (1940). Hornbill studies. *Ibis* (14), **4**, 639–56.

Morgan, C. Ll. (1930). *The Animal Mind.* London.

Morley, A. (1936). The winter behaviour of moor-hens. *Brit. Birds*, **30**, 122–4.

Morley, A. (1941). The behaviour of a group of resident British starlings (*Sturnus v. vulgaris* Linn.). *Naturalist*, **788**, 55–61.

Moseley, H. N. (1879). *Notes by a Naturalist on the Challenger.* London.

Müller, A. and Müller, K. (1822–23). *Thiere der Heimath.* Kassel and Berlin.

Murchison, C. (1935). The experimental measurement of a social hierarchy in *Gallus domesticus. J. soc. Psychol.* **6**, 3–30.

Murchison, C. (1935). The experimental measurement of a social hierarchy in *Gallus domesticus.* I. The direct identification and measurement of social reflex no. 1 and social reflex no. 2. *J. genet. Psychol.* **12**, 3-39.

Murphy, R. C. (1936). *Oceanic Birds of South America.* New York.

Naumann, J. A. and Naumann, J. F. (1822–53). *Naturgeschichte der Vögel Deutschlands.* Leipzig.

Neff, J. A. (1937). Nesting distribution of the tri-colored redwing. *Condor*, **39**, 61–81.

Nelson, A. (1887). *Report upon Natural History Collections made in Alaska.* Washington, D.C.

Nethersole-Thompson, C. and Nethersole-Thompson, D. (1939). Some observations on the sexual life, display, and breeding of the red grouse as observed in Inverness-shire. *Brit. Birds*, **32**, 247–54.

Newton, A. (1896). *Dictionary of Birds*. London.

Nice, M. M. (1929). Observations on the nesting of a pair of yellow-crowned night-herons. *Auk*, **46**, 170–6.

Nice, M. M. (1936). The nest in the rose-hedge. *Bird-Lore*, **38**, 337–43.

Nice, M. M. (1937). Studies in the life history of the song sparrow. I. A population study of the song sparrow. *Trans. Linn. Soc. N.Y.* **4**, 1–247.

Nice, M. M. (1938). Territory and mating with the song sparrow. *Proc. 8th Int. orn. Congr. Oxford*, 1934, pp. 324–38. Oxford.

Nice, M. M. (1939). The social kumpan and the song sparrow. *Auk*, **56**, 255–62.

Nicholson, E. M. (1927). *How Birds Live*. London.

Nicholson, E. M. (1929). Cited in J. M. Winterbottom, Communal display in birds. *Proc. zool. Soc. Lond.*, p. 192.

Nicholson, E. M. (1931). Field notes on the Guiana king humming bird. *Ibis* (13), **I**, 534–53.

Niethammer, G. (1937–38). *Handbuch der deutschen Vogelkunde*. Leipzig.

Nissen, H. W. (1931). A field study of the chimpanzee; observations of chimpanzee behavior and environment in western French Guinea. *Comp. Psychol. Monogr.* **8**.

Noble, G. K. (1936). Courtship and sexual selection of the flicker (*Colaptes auratus luteus*). *Auk*, **53**, 269–82.

Noble, G. K. (1938). Sexual selection among fishes. *Biol. Rev.* **13**, 133–58.

Noble, G. K. (1939). The rôle of dominance in the social life of birds. *Auk*, **56**, 263–73.

Noble, G. K. and Borne, R. (1938). The social hierarchy in *Xiphophorus* and other fishes. *Bull. Ecol. Soc. Amer.* **19**, 14.

Noble, G. K. and Bradley, H. T. (1933). The mating behavior of lizards; its bearing on the theory of natural selection. *Ann. N.Y. Acad. Sci.* **35**, 25–100.

Noble, G. K. and Vogt, W. (1935). An experimental study of sex recognition in birds. *Auk*, **52**, 278–86.

Noble, G. K. and Wurm, M. (1940). The effect of hormones on the breeding of the laughing gull. *Anat. Rec.* **78**, suppl. 25.

Noble, G. K. and Wurm, M. (1940a). The effect of testosterone propionate on the black-crowned night heron. *Endocrinology*, **26**, 837–50.

Noble, G. K., Wurm, M. and Schmidt, A. (1938). The social behavior of the black-crowned night heron. *Auk*, **55**, 7–40.

Nutting, C. C. (1883). On a collection of birds from Nicaragua. *Proc. U.S. nat. Mus.* **6**, 384–5.

Ogilvie-Grant, W. R. (1896). *Game Birds*. London.

348   BIBLIOGRAPHY

Ogilvie-Grant, W. R. (1914). Cited in W. P. Pycraft, *Courtship of Animals*. London.
Oldham, C. (1938). Cited in H. F. Witherby *et al.*, *Handbook of British Birds*, 1, 59.
Olina, G. P. (1622). *Uccelliera, overo Discorso della Natura e Propriete de Diversi Uccelli, e in Particularo di que che Cantano, con il Modo di Prendergli, Conoscergli, Allivargli e Mantenergli*, 1. Rome.
Osmaston, B. B. (1934). Robin with four broods. *Brit. Birds*, 28, 113–15.
Osmaston, B. B. (1941). Duetting in birds. *Ibis* (14), 5, 310–11.

Palmgren, P. (1934). Balz als Ausdruck der Zugektase bei einem gekäfigten Fitislaubsänger. *Ornis fenn.* 4, 84–6.
Pavlov, I. P. (1927). *Conditioned Reflexes*. Oxford.
Pavlov, I. P. (1929). *Lectures on Conditioned Reflexes*. London.
Peckham, G. W. and Peckham, E. G. (1889). Observations on sexual selection in spiders of the family Attidae. *Occ. Pap. nat. Hist. Soc. Wis.* 1, 1–60.
Peckham, G. W. and Peckham, E. G. (1890). Additional observations on sexual selection in spiders of the family Attidae. *Occ. Pap. nat. Hist. Soc. Wis.* 2, 115–51.
Peitzmeier, J. (1936). Die Akinese bei Vögeln ein Instinkt? *Orn. Mber.* 44, 110–16.
Perry, J. C. (1938). The influence of diet on the gonad activity of the English sparrow, *Passer domesticus* Linnaeus. *Proc. Soc. exp. Biol., N.Y.*, 38, 716–19.
Perry, R. B. (1938). *At the Turn of the Tide*. London.
Perry, R. B. (1940). *Lundy: Isle of Puffins*. London.
Peters, J. L. (1931–40). *Check List of Birds of the World*. Cambridge, Mass.: Harvard Univ. Press.
Petronius. *Satyricon*, LV.
Pettingill, O. S. (1937). Behavior of black skimmers at Cardwell Island, Virginia. *Auk*, 54, 237–44.
Phillips, J. C. (1922–26). *A Natural History of the Ducks*. Boston and New York.
Phillips, R. (1911). Further notes on the regent bird, *Sericulus melinus*. *Avicult. Mag.* (3), 2, 355–7.
Pickwell, G. and Smith, E. (1938). The Texas night-hawk in its summer home. *Condor*, 40, 193–215.
Pieters, D. (1935). De Argusfazant. *Tropische Natuur*, 24, 8–15.
Pinchot, G. (1931). *To the South Seas*. London.
Pitman, C. R. S. (1912). The painted snipe (*Rostratula capensis*). *J. Bombay nat. Hist. Soc.* 21, 666–7.

Pitt, F. (1927). *Animal Mind.* London.

Pitt, F. (1931). *The Intelligence of Animals: Studies in Comparative Psychology.* London.

Pliny, C. P. S., the elder. *Naturalis Historia.*

Poll, H. (1911). Über Vogelmischlinge. *Verh. V. Int. orn. Kongr. Berlin,* 1910, pp. 399–468. Berlin.

Ponting, H. G. (1921). *The Great White South.* London.

Portielje, A. F. J. (1926). Zur Ethologie bezw. Psychologie von *Botaurus stellaris. Ardea,* **15**, 1–15.

Portielje, A. F. J. (1927). Zur Ethologie bezw. Psychologie von *Phalacrocorax carbo subcormorans* (Brehm). *Ardea,* **16,** 107–23.

Portielje, A. F. J. (1928). Zur Ethologie bezw. Psychologie der Silbermöwe *Larus argentatus argentatus* Pont. *Ardea,* **17,** 112–49.

Portielje, A. F. J. (1931). Versuch zu einer verhaltungspsychologischen Deutung des Balz-Gebarens der Kampfschnepfe. *Proc. 7th Int. orn. Congr. Amsterdam,* 1930, pp. 156–72. Amsterdam.

Poulton, E. B. (1890). *The Colours of Animals, Their Meaning and Use, especially considered in the case of Insects.* London.

Pratt, J. P. (1932). Endocrine disorders in sex function in man. Chap. xix in E. Allen, *Sex and Internal Secretions.* Baltimore.

Priestley, M. (1937). *A Book of Birds.* London.

Putzig, P. (1937). Von der Beziehung des Zugablaufs zum Inkretdrüsensystem. *Vogelzug,* **8,** 116–30.

Putzig, P. (1939). Sechste Rückmeldung einer kastrierten Nebelkrähe (*Corvus corvix*) in Heimzugrichtung. *Vogelzug,* **10,** 171–2.

Pycraft, W. P. (1913). Cited in F. B. Kirkman, *The British Bird Book,* **4,** 233, 267.

Pycraft, W. P. (1914). *The Courtship of Animals.* 2nd ed. London.

Pycraft, W. P. (1934). *Birds of Great Britain and their Natural History.* London.

Ramsay, J. S. P. (1934). Cited in A. H. Chisholm, *Bird Wonders of Australia.* Sydney.

Rand, A. L. (1938). On the breeding habits of some birds of paradise in the wild. Results of the Archbold Expeditions. No. 22. *Amer. Mus. Novit.* **993,** 1–8.

Rand, A. L. (1940). Breeding habits of the birds of paradise: *Macgregoria* and *Diphyllodes.* Results of the Archbold Expeditions. No. 26. *Amer. Mus. Novit.* **1073,** 1–14.

Rand, A. L. (1940 a). Courtship of the magnificent bird of paradise. *Nat. Hist. Mag., N.Y.* **45,** No. 3.

Rand, A. L. (1941). Courtship feeding in birds. *Auk,* **58,** 57–9.

Rankin, M. N. and Rankin, D. H. (1940). The breeding behaviour of the Irish dipper. *Irish Nat. J.* **7**, 273–82.

Raspail, X. (1901). Cérémonie de secondes noces chez les Garruliens *Pica pica* sp. et *Garrulus gl. glandarius* (L.). *Bull. Soc. zool. Fr.* **26**, 104–9.

Rawlings, M. K. (1938). *The Yearling.* New York.

Report of Wild Birds Protection Committee of Norfolk Naturalists' Trust (1938).

Rettig, A. (1927). *Beitr. FortPflBiol. Vögel,* **3**, 102.

Richards, O. W. (1927). Sexual selection and allied problems in the insects. *Biol. Rev.* **2**, 298–360.

Richmond, W. K. (1939). On the strange courtships of British mergansers. *Naturalist,* **993**, 267–76.

Richter, R. (1939). Weitere Beobachtungen an einer gemischten Kolonie von *Larus fuscus graellsi* Brehm und *Larus argentatus* Pontopp. *J. Orn.* **87**, 75–86.

Rickman, P. (1931). *A Bird-Painter's Notebook.* London.

Riddle, O. (1935). The lactogenic factor of the pituitary. *J. Amer. med. Ass.* **104**, 636–7.

Riddle, O., Bates, R. W. and Lahr, E. L. (1935). Prolactin induces broodiness in fowl. *Amer. J. Physiol.* **111**, 352–60.

Riddle, O., Bates, R. W. and Lahr, E. L. (1937). Cf. Bates, Riddle and Lahr (1937).

Riddle, O. and Braucher, P. F. (1931). Studies on the physiology of reproductions in birds. XXX. Control of the special secretion of the crop gland in pigeons by an anterior pituitary hormone. *Amer. J. Physiol.* **97**, 617–25.

Riddle, O. and Dykshorn, S. W. (1932). The secretion of crop milk in the castrate pigeon. *Proc. Soc. exp. Biol. N.Y.,* **29**, 1213.

Ritter, W. E. (1927). *The Natural History of our Conduct.* New York.

Roberts, B. B. (1934). Notes on the birds of central and south-east Iceland, with special reference to food-habits. *Ibis* (13), **4**, 239–64.

Roberts, B. B. (1940 *a*). The breeding behaviour of penguins with special reference to *Pygoscelis papua* (Forster). *British Graham Land Expedition, 1934–7, Scientific Reports,* **I**, 195–254.

Roberts, B. B. (1940 *b*). The life cycle of Wilson's petrel, *Oceanites oceanicus* (Kuhl). *British Graham Land Expedition, 1934–7, Scientific Reports,* **I**, 141–94.

Robinson, H. W. (1935). Gannets wintering on the Bass Rock. *Scot. nat.* **213**, 78.

Rohweder, J. (1891). Am Balzplatz von *Gallinago major. J. Orn.* **19**, 419–26.

Rothmann, M. and Teuber, E. (1915). Einzelausgabe aus der Anthropoidenstation auf Teneriffa. I. Ziele und Aufgaben der Station sowie erste Beobachtungen an den auf ihr gehaltenen Schimpansen. *Abh. preuss. Akad. Wiss.* pp. 1–20.

Rothschild, M. (1941). Personal communication.

Rothschild, W. L. (1893–1900). *The Avifauna of Laysan and the Neighbouring Islands.* London.

Rowan, W. (1925). Relation of light to migration and developmental changes. *Nature, Lond.,* 115, 494–5.

Rowan, W. (1926). On photoperiodism, reproductive periodicity, and the annual migration of birds and certain fishes. *Proc. Boston Soc. nat. Hist.* 38, 147.

Rowan, W. (1927). Migration and reproductive rhythm in birds. *Nature, Lond.,* 119, 351–2.

Rowan, W. (1927 a). Notes on Alberta waders included in the British list. V. *Tryngites subruficollis,* buff-breasted sandpiper. *Brit. Birds,* 20, 186–92.

Rowan, W. (1928). Reproductive rhythms in birds. *Nature, Lond.,* 122, 11–12.

Rowan, W. (1930). Experiments in bird migration. II. Reversed migration. *Proc. nat. Acad. Sci., Wash.,* 16, 520–5.

Rowan, W. (1931). *The Riddle of Migration.* Baltimore.

Rowan, W. (1937). Effects of traffic disturbance and night illumination on London starlings. *Nature, Lond.,* 139, 668–9.

Rowan, W. (1938). Light and seasonal reproduction in animals. *Biol. Rev.* 13, 374–402.

Rowan, W. and Batrawi, A. M. (1939). Comments on the gonads of some European migrants collected in East Africa immediately before their spring departure. *Ibis* (14), 3, 58–65.

Rudolph, Crown Prince (1898). Cited in K. Groos, *The Play of Animals.* New York.

Russell, E. S. (1938). *The Behaviour of Animals: An Introduction to its Study.* 2nd ed. London.

Ryves, B. H. (1929). Three redshanks at one nest. *Brit. Birds,* 23, 103.

Sachs, C. (1938). *World History of the Dance,* tr. B. Schönberg. London.

Salvin, O. (1873). On the tail feathers of *Momotus. Proc. zool. Soc. Lond.* pp. 429–33.

Savory, T. H. (1928). *The Biology of Spiders.* London.

Saunders, A. A. (1937). Injury-feigning by a wood duck. *Auk,* 54, 202.

Saxby, H. L. (1874). *The Birds of Shetland.* Edinburgh and London.

Schenk, J. (1929). Die Brutinvasion des Rosenstares in Ungarn im Jahre 1925. *Verh. VI. Int. orn. Kongr. Köpenhagen,* 1926, pp. 249–64. Berlin.

Schildmacher, H. (1933). Zur Physiologie des Zugtriebes. I. Versuche mit weiblichem Sexualhormon. *Vogelzug,* 4, 21–4.

Schildmacher, H. (1934). Zur Physiologie des Zugtriebes. II. Weitere Versuche mit weiblichem Sexualhormon. *Vogelzug*, **5**, 1–9.

Schildmacher, H. (1937). Zur Physiologie des Zugtriebes. III. Versuche mit künstlich verlängerten Tagesdauer. *Vogelzug*, **8**, 107–14.

Schildmacher, H. (1938). Zur Physiologie des Zugtriebes. IV. Weitere Versuche mit künstlich veränderten Belichtungszeit. *Vogelzug*, **9**, 146–52.

Schjelderup-Ebbe, T. (1922*a*). Beiträge zur Sozialpsychologie des Haushuhns. *Z. Psychol.* **88**, 225–52.

Schjelderup-Ebbe, T. (1922*b*). Soziale Verhaltnisse bei Vögeln. *Z. Psychol.* **90**, 106–7.

Schjelderup-Ebbe, T. (1923*a*). Weitere Beiträge zur Sozial- und Individualpsychologie des Haushuhns. *Z. Psychol.* **92**, 60–87.

Schjelderup-Ebbe, T. (1923*b*). Das Leben der Wildente in der Zeit der Paarung. *Psychol. Forsch.* **3**, 12–17.

Schjelderup-Ebbe, T. (1924*a*). Zur Sozialpsychologie der Vögel. *Z. Psychol.* **95**, 36–84.

Schjelderup-Ebbe, T. (1924*b*). Fortgesetzte biologische Beobachtungen des *Gallus domesticus*. *Psychol. Forsch.* **5**, 343–5.

Schjelderup-Ebbe, T. (1925). Soziale Verhältnisse bei Säugetieren. *Z. Psychol.* **97**, 145.

Schjelderup-Ebbe, T. (1930). Psychologische Beobachtungen an Vögeln. *Z. allg. Psychol.* **25**, 363–6.

Schjelderup-Ebbe, T. (1931). Die Despotie im sozialen Leben der Vögel. In Richard Thurnvald, *Forsch. Völkerpsych. Soziol.* **10**, 77–137.

Schjelderup-Ebbe, T. (1935*a*). Social behavior in birds. Chap. xx in C. Murchison, *Handbook of Social Psychology*, pp. 947–72. Clark Univ. Press.

Schjelderup-Ebbe, T. (1935*b*). Despotism amongst birds. *Scand. sci. Rev.* **3**, nos. 3 and 4, pp. 10–82.

Schmid, B. (1936). *Interviewing Animals*. London, Eng. tr. *Begegnung mit Tieren* (1935). Munich.

Schomburgk, R. H. (1841). Journey from Fort San Joaquim, on the Rio Branco, to Roraima, and thence by the Rivers Parima and Merewari to Esmeralda, on the Orinoco in 1838–9. *J. roy. Geogr. Soc.* **10**, 191–247.

Schwenter, D. (1636). *Deliciae Physicomathematicae*. Nürnberg.

Seebohm, H. (1885). *History of British Birds*. London.

Seigne, J. W. and Keith, E. C. (1936). *Woodcock and Snipe*. London.

Selous, E. (1901). *Bird Watching*. London.

Selous, E. (1901–2). An observational diary of the habits, mostly domestic, of the great crested grebe and the peewit. *Zoologist* (4), **5**, 161–83, 339–50, 454–62; **6**, 133–44.

Selous, E. (1902). Note on the pairing of moorhens. *Zoologist* (4), **6**, 196–7.

Selous, E. (1905 a). *Bird-Life Glimpses*. London.

Selous, E. (1905 b). *The Bird-Watcher in the Shetlands; with some Notes on Seals—and digressions*. London.

Selous, E. (1906–7). Observations tending to throw light on the question of sexual selection in birds, including a day-to-day diary on the breeding habits of the ruff (*Machetes pugnax*). *Zoologist* (4), **10**, 201–19, 285–94, 419–28; **11**, 60–5, 161–82, 367–80.

Selous, E. (1909–10). An observational diary on the nuptial habits of the blackcock (*Tetrao tetrix*) in Scandinavia and England. *Zoologist* (4), **13**, 400–13; **14**, 23–9, 51–6, 176–82, 248–65.

Selous, E. (1913–16). A diary of ornithological observations made in Iceland. *Zoologist* (4), **17**, 57–65 (eagles), 92–104 (eagles and swans), 129–36 (swans), 294–313 (swans and merlins), 409–22 (divers and falcons); **18**, 63–74 (grebes), 213–25 (grebes and rooks); **19**, 58–66, 169–74 (merlins), 303–7 (phalaropes); **20**, 54–68, 139–52 (phalaropes), 267–72 (divers).

Selous, E. (1914). The earlier breeding habits of the red-throated diver. *Wild Life*, **3**, 206–13.

Selous, E. (1916). On the sexual origin of the nidificatory, incubatory and courting display instincts in birds. An answer to criticism. *Zoologist* (4), **20**, 401–12.

Selous, E. (1927). *Realities of Bird Life*. London.

Selous, E. (1931). *Thought-transference (or what?) in Birds*. London.

Selous, E. (1933). *The Evolution of Habit in Birds*. London.

Semper, K. (1873). *Die Palau-Inseln*. Leipsig.

Sergi, G. (1901). *Les Émotions*. Fr. tr. London.

Seth-Smith, D. (1907). The importance of agriculture as an aid to the study of ornithology. *Proc. 4th Int. orn. Congr. London*, 1905. *Ornis*, **14**, 663–75.

Seth-Smith, D. (1914). Notes from the Zoological Gardens. *Wild Life*, **4**, 54.

Seth-Smith, D. (1923). On the display of the magnificent bird of paradise. *Proc. zool. Soc. Lond.* pp. 609–13.

Sharpe, R. B. and Dresser, H. E. (1871–81). *A History of the Birds of Europe, including all the species inhabiting the Western Palaearctic Region*. London.

Shelford, R. W. C. (1916). *A Naturalist in Borneo*, ed. E. B. Poulton. London.

Shelley, G. E. (1905). *Birds of Africa*, **4**, 131. London.

Sherrington, C. (1940). *Man on his Nature*. Cambridge.

Shoemaker, H. H. (1939 a). Effect of testosterone propionate on behavior of the female canary. *Proc. Soc. exp. Biol., N.Y.*, **41**, 299–302.

Shoemaker, H. H. (1939 b). Social hierarchy in flocks of the canary. *Auk*, **56**, 381–406.

Sick, H. (1936). Vom Balzflug des Schwarzen Milans. *Beitr. FortPflBiol. Vögel*, **12**, 188–9.

Siddall, C. K. (1910). Willow wren feigning injury. *Brit. Birds*, **4**, 118.

Sieber, H. (1932). Beobachtungen über die Biologie des Kranichs (*Megalornis gr. grus*). *Beitr. FortPflBiol. Vögel*, **8**, 134–9, 176–80.

Simon, J. R. (1940). Mating performance of the sage grouse. *Auk*, **57**, 467–71.

Simson, C. C. (1907). On the habits of the birds of paradise and bower birds of British New Guinea. *Ibis* (9), **1**, 380–7.

Skard, A. G. (1936). Studies in the psychology of needs. Observations and experiments on the sexual needs in hens. *Acta Psychol. Kbh.* **2**, 175–232.

Skutch, A. F. (1935). Helpers at the nest. *Auk*, **52**, 257–73.

Skutch, A. F. (1937). Life history of the black-chinned jacamar. *Auk*, **54**, 135–46.

Skutch, A. F. (1940). Rieffer's hummingbird. In A. C. Bent, Life histories of North American cuckoos, goatsuckers, hummingbirds and their allies. *Smithsonian Inst. U.S. nat. Mus. Bull.* **176**, 432–43.

Söderberg, R. (1929). The genesis of the decorative and building instincts of bower birds. *Verh. VI. Int. orn. Kongr. Köpenhagen, 1926*, pp. 297–337. Berlin.

Sokolowsky, A. (1923). The sexual life of the great apes. *Urol. cutan. Rev.* **27**, 612–15.

Southern, H. N. (1938). Posturing and related activities of the common tern (*Sterna h. hirundo* L.). *Proc. zool. Soc. Lond.* A, **108**, 423–31.

Southern, H. N. and Lewis, W. A. S. (1938). The breeding behaviour of Temminck's stint. *Brit. Birds*, **31**, 314–21.

Southern, H. N. and Venables, L. S. V. (1936). The courtship of the bonxie. *Field*, **168**, 98.

Spennemann, A. (1928). Zur Brutbiologie von *Centropus javanicus* (Dumont). *Beitr. FortPflBiol. Vögel*, **4**, 139–44.

Stadler, H. (1932). La voix des Chouettes de l'Europe Moyenne. *Alauda*, **4**, 174–91, 271–83, 407–15.

Steinbacher, G. (1931). Brutbiologie der Silbermöwe und Brandseeschwalbe. *J. Orn.* **79**, 349–53.

Steinbacher, G. (1938). Über einige brutbiologische Beobachtungen im Berliner Zoologischen Garten im Jahre 1937. *Beitr. FortPflBiol. Vögel*, **14**, 55.

Steiniger, F. (1937). Die biologische Bedeutung der 'tierischen Hypnose' bei Vögeln. *J. Orn.* **85**, 593–603.

Steinmetz, H. (1931). Brutbiologisches vom Mönchsittich. *Zool. Gart., Lpz.*, **4**, 140–53.

Sterne, L. (1768). *A Sentimental Journey*. London.

Stimmelmayr, A. (1932). Neue Wege zur Erforschung des Vogelzuges. *Verh. orn. Ges. Bayern*, **19**, 418–46.

Stonor, C. R. (1936). The evolution and mutual relationship of some members of the Paradiseidae. *Proc. zool. Soc. Lond.* pp. 1177–85.

Stonor, C. R. (1938). Some features of the variation of the birds of paradise. *Proc. zool. Soc. Lond.* B, **108**, 417–81.

Stonor, C. R. (1939). Notes on the breeding habits of the common screamer (*Chauna torquata*). *Ibis* (14), **3**, 45–9.

Stonor, C. R. (1940). *Courtship and Display among Birds.* London.

Storrs, R. (1937). *Orientations.* London.

Stow, G. W. and Theal, G. M. (1905). *The Native Races of South Africa.* London.

Stresemann, E. (1928). *Aves*, in Kukenthal, *Handbuch der Zoologie*, **7**, 404–15.

Strijbos, J. P. (1935). Nesting of the black-winged stilt. *Thijse-Gedenkboek*, pp. 123–31.

Strong, R. M. (1939). A bibliography of birds, with special reference to anatomy, behavior, bio-chemistry, embryology, pathology, physiology, genetics, ecology, agriculture, evolution and related subjects. *Zool. Ser. Field Mus. Nat. Hist.* **25**, 1–464, 465–937.

Stubbs, F. J. (1910). Ceremonial gatherings of the magpie. *Brit. Birds*, **3**, 334–6.

Swarth, H. S. (1935). Injury-feigning in nesting birds. *Auk*, **52**, 352–4.

Taczanowski, L. and Stolzmann, J. (1881). Notice sur la *Loddigesia mirabilis* (Boure.). *Proc. zool. Soc. Lond.* pp. 827–34.

Taverner, P. A. (1936). Injury feigning by birds. *Auk*, **53**, 366.

Tavistock, Marquess of (1936). Cited in F. H. A. Marshall, Sexual periodicity and the causes which determine it. *Philos. Trans. roy. Soc.* B **226**, 445.

Thomas, W. B. (1939). Note on brush turkeys. *Spectator*, **162**, 806.

Thompson, D'Arcy W. (1910). *The Works of Aristotle*, **4**. *Historia Animalium.* Oxford.

Thompson, W. H. (1938). Cited in H. F. Witherby *et al.*, *Handbook of British Birds*, **2**, 293.

Thomson, D. F. (1934). Some adaptations for the disposal of faeces. The hygiene of the nest in Australian birds. *Proc. zool. Soc. Lond.* pp. 701–6.

Thomson, J. A. (1932). *Scientific Riddles.* London.

Thorpe, W. H. (1938). Further experiments on olfactory conditioning in a parasitic insect. The nature of the conditioning process. *Proc. roy. Soc.* B, **126**, 370–97.

Thorpe, W. H. (1939). Further studies in pre-imaginal olfactory conditioning in insects. *Proc. roy. Soc.* B, **127**, 424–33.

Times, The (1938*a*). 30 July. Note on chimpanzees.

Times, The (1938*b*). 6 August. Note on chimpanzees.

Tinbergen, N. (1931). Zur Paarungsbiologie der Fluszseeschwalbe (*Sterna hirundo hirundo* L.). *Ardea*, **20**, 1–18.

Tinbergen, N. (1932). Vergelijkende waarnemingen aan enkele Meeuwen en Sterns. *Ardea*, **21**, 1–13.

Tinbergen, N. (1935). Field observations of East Greenland birds. I. The behaviour of the red-necked phalarope (*Phalaropus lobatus* L.) in spring. *Ardea*, **24**, 1–42.

Tinbergen, N. (1936*a*). Zur Soziologie der Silbermöwe, *Larus a. argentatus*. *Beitr. FortPflBiol. Vögel*, **12**, 89–96.

Tinbergen, N. (1936*b*). The function of sexual fighting in birds; and the problem of the origin of territory. *Bird Banding*, **7**, 1–8.

Tinbergen, N. (1939*a*). Field observations of East Greenland birds. II. The behavior of the snow bunting (*Plectrophenax nivalis subnivalis* (Brehm)) in spring. *Trans. Linn. Soc. N.Y.*, pp. 1–94.

Tinbergen, N. (1939*b*). Social organization among vertebrates. *Amer. Midl. Nat.* **21**, 225.

Tinbergen, N. and Kuenen, D. J. (1939). Über die auslösenden und die richtunggebenden Reizsituationen der Sperrbewegung von jungen Drosseln (*Turdus m. merula* L. und *T. e. ericetorum* Turton). *J. Tierpsychol.* **3**, 37–60.

Tomlinson, H. J. (1930). A study in instinct. *Windsor Mag.* Aug.

Tongue, H. (1909). *Bushman Paintings*. Oxford.

Townsend, C. W. (1911). The courtship and migration of the red-breasted merganser. *Auk*, **28**, 341–5.

Townsend, C. W. (1913). *Sand Dunes and Salt Marshes*. Boston.

Townsend, C. W. (1928). The song of the green heron (*Butorides v. virescens*). *Auk*, **45**, 498–99.

Treatt, C. C. (1930). *Out of the Beaten Track*. London.

Tregallas, T. (1931). The truth about the lyre bird. *Emu*, **30**, 243–50.

Turner, E. L. (1914). The nightjar in Sussex. *Wild Life*, **3**, 16–24.

Turner, E. L. (1924). *Broadland Birds*. London.

Turner, E. L. (1928). *Bird Watching on Scolt Head*. London.

Uhrich, J. (1938). The social hierarchy in albino mice. *J. comp. Psychol.* **25**, 373–413.

van Oordt, G. J. and Bruyns, M. F. M. (1938). Die Gonaden übersommernder Austernfischer (*Haematopus ostralegus* L.). *Z. Morph. ökol. Tiere*, **34**, 161–72.

van Someren, V. D. (1933). Some observations on the nesting habits of the blackbird. *Scot. Nat.* **201**, 75–84.

Venables, L. S. V. (1938). The nesting of the nuthatch. *Brit. Birds*, **32**, 26–33.
Venables, L. S. V. (1940). The nesting behaviour of the Galapagos mocking-bird. *Ibis* (14), **4**, 629–39.
Venables, L. S. V. and Lack, D. L. (1936). Further notes on territory in the great crested grebe. *Brit. Birds*, **30**, 60–9.
Verlaine, L. (1934). L'instinct et l'intelligence chez les oiseaux. *Recherches Philosoph.* **3**, 285–305.
Verrill, A. H. (1938). *Strange Birds and their Stories*. London.
Verwey, J. (1929). Die Paarungsbiologie des Fischreihers (*Ardea cinerea* L.). *Verh. VI. Int. orn. Kongr. Kopenhagen*, 1926, pp. 390–413. Berlin.
Verwey, J. (1930). Die Paarungsbiologie des Fischreihers. *Zool. Jb.*, Abt. allg. Zool. Physiol. **48**, 1–120.
Vetulani, T. (1931). Untersuchungen über das Wachstum der Säugetiere in Abhängigkeit von der Anzahl zusammengeschaltener Tiere. *Biol. generalis*, **7**, 71–98.
Vincent, J. (1938). Letter on water rail's behaviour. *Country Life*, 12 Feb.
von Frisch, K. (1937). The language of bees. *Sci. Progr.* **32**, 29–37.
von Fürer-Haimendorf, C. (1938). Spring festival among the Konyak Nagas of Assam. *Geogr. Mag.* **7**, 248.
von Holst, E. (1935). Alles oder Nichts, Block, Alternans, Bigemini und verwandte Phänomene als Eigenschaften des Rückenmarks. *Pflüg. Arch. ges. Physiol.* **236**, 149–59.
von Holst, E. (1936a). Vom Dualismus der motorischen und der automatisch-rhythmischen Funktion im Rückenmark und vom Wesen des automatischen Rhythmus. *Pflüg. Arch. ges. Physiol.* **237**, 356–78.
von Holst, E. (1936b). Versuchen zur relativen Koordination. *Pflüg. Arch. ges. Physiol.* **237**, 93–122.
von Holst, E. (1937a). Über den 'Magnet-Effekt' als koordinierendes Prinzip im Rückenmark. *Pflüg. Arch. ges. Physiol.* **237**, 655–82.
von Holst, E. (1937b). Regulationsfähigkeit im Zentralnervensystem. *Naturwissenschaften*, **25**, 625–31, 641–7.
von Ihering, H. (1885). Die Vögel der Umgegend von Taquara do Mundo, Prov. Rio Grande do Sul. *Z. ges. Orn.* **2**, 138.
von Uexküll, J. (1909). *Umwelt und Innenwelt der Tiere*. Berlin.
von Vormann, P. F. (1911). Tänze und Tanzfestlichkeiten der Monumbo-Papua. *Anthropos*, **6**, 411–27.

Wachs, H. (1933). Paarungsspiele als Artcharactere, Beobachtungen an Möwen und Seeschwalben. *Verh. dtsch. zool. Ges.* pp. 192–202.
Wagner, H. O. (1938). Beobachtungen über die Balz des Paradiesvogels *Paradisaea gulielmi* Cab. *J. Orn.* **86**, 550–3.
Wall, F. (1921). *Ophidia Taprobanica, or The Snakes of Ceylon*. Colombo.

Wallace, A. R. (1869). *The Malay Archipelago.*. London.

Wallace, G. J. (1939). Bicknell's thrush; its taxonomy, distribution and life history. *Proc. Boston Soc. nat. Hist.* **41**, 211–402.

Warner, L. H. (1927). A study of sex behavior in the white rat by means of the obstruction method. *Comp. Psychol. Monogr.* **4**.

Warner, L. H. (1928). A study of hunger behavior in the white rat by means of the obstruction method: comparison of sex and hunger behaviour. *J. comp. Psychol.* **8**, 273–99.

Waterhouse, J. D. (1939). The variegated wren. *Emu*, **39**, 93–4.

Watson, J. B. (1908). The behavior of noddy and sooty terns. *Pap. Tortugas Lab., Publ. Carneg. Instn.* **103**, 187–225.

Weir, J. (1871). Cited in C. Darwin, *Descent of Man*. London.

Wetmore, A. (1926). Observations on the birds of Argentina, Paraguay and Chile. *Smithsonian Inst. Bull. U.S. nat. Mus.* **133**, 1–448.

Wetmore, A. (1940). A systematic classification for the birds of the world. Revised and amended. *Smithsonian Misc. Coll.* **99**, no. 7, pp. 1–11.

Wetmore, A. and Swales, B. H. (1931). The birds of Haiti and the Dominican Republic. *U.S. nat. Mus. Bull.* **155**, 345–52.

Whetham, E. O. (1933). Factors modifying egg production with special reference to seasonal changes. *J. agric. Sci.* **23**, 383–418.

White, G. (1789). *The Natural History and Antiquities of Selborne, in the County of Southampton*. London.

Whitman, C. O. (1919). The behavior of pigeons. Posthumous works of C. O. Whitman, **3**, 1–161. Ed. by H. A. Carr. *Publ. Carneg. Instn*, **257**.

Wilkes, A. H. P. (1928). The birds of the region south of Lake Nyasa. I. Non-passerine birds. *Ibis* (12), **4**, 698–99.

Wilkins, G. H. (1923). Report on the birds collected during the voyage of the *Quest* (Shackleton-Rowett expedition) to the southern Atlantic. *Ibis* (11), **5**, 474–511.

Williamson, K. (1937). *The Sky's their Highway*. London.

Williamson, K. (1941). First brood of swallows assisting to feed second brood. *Brit. Birds*, **34**, 221.

Wilson, E. A. (1907). *British National Antarctic Expedition*, **2**. London.

Winterbottom, J. M. (1928). The display of Wilson's bird of paradise, *Schlegelia wilsoni*. *Ibis* (12), **4**, 318–20.

Winterbottom, J. M. (1929). Studies in sexual phenomena. VI. Communal display in birds. *Proc. zool. Soc. Lond.* pp. 189–95.

Witherby, H. F. (1936). 'Injury-feigning' by wood-lark. *Brit. Birds*, **30**, 81.

Witherby, H. F., Jourdain, F. C. R., Ticehurst, N. F. and Tucker, B. W. (1938) Vols. 1 and 2; (1939) Vol. 3; (1940) Vol. 4; (1941) Vol. 5. *The Handbook of British Birds*. London.

Witschi, E (1935). Seasonal sex characters in birds and their hormonal control. *Wilson Bull.* **47**, 177–88.

Witschi, E. and Miller, R. A. (1938). Ambisexuality in the female starling. *J. exp. Zool.* **79**, 475–87.

Wolfson, A. (1940). A preliminary report on some experiments on bird migration. *Condor,* **42**, 93.

Wood, R. C. (1940). 'Drumming' of the black-throated honey-guide (*Indicator indicator*). *Ostrich,* **11**, 50–1.

Woods, R. S. (1927). The humming birds of California. *Auk,* **44**, 297–318.

Woods, R. S. (1940). Anna's humming bird. In A. C. Bent, Life histories of North American cuckoos, goatsuckers, humming birds and their allies. *Smithsonian Inst. Bull. U.S. nat. Mus.* **176**, 371–87.

Woodward, M. (1938). *Adventures in Woodcraft.* London.

Wundt, W. (1912). *Lectures on Human and Animal Psychology*, Eng. tr. London.

Wynne-Edwards, V. C. (1930). On the waking-time of the nightjar (*Caprimulgus e. europaeus*). *J. exp. Biol.* **7**, 241–7.

Wynne-Edwards, V. C. (1939). Intermittent breeding of the fulmar (*Fulmarus glacialis* L.) with some general observations on non-breeding in sea-birds. *Proc. zool. Soc. Lond.* **109**, 127–32.

Yamashina, Marquis (1938). A sociable breeding habit among Timaliine birds. *Report 9th Int. orn. Congr., Rouen,* 1938, pp. 453–6.

Yarrell, W. (1843). *A History of British Birds.* London.

Yeates, G. K. (1934). *The Life of the Rook.* London.

Yeates, G. K. (1936). On the fighting of blackcock. *Brit. Birds,* **30**, 34–7.

Yeates, G. K. (1940). Some notes on the bittern. *Brit. Birds,* **34**, 98.

Yeates, G. K. (1941). Strange behaviour of long-eared owls in a thunderstorm. *Brit. Birds,* **35**, 82.

Yerkes, R. M. (1925). *Almost Human.* London.

Yerkes, R. M. and Yerkes, A. W. (1929). *The Great Apes: A Study of Anthropoid Life.* Yale and London.

Yerkes, R. M. and Yerkes, A. W. (1935). Social behavior in infrahuman primates. Chap. xxi in C. Murchison, *Handbook of Social Psychology,* pp. 973–1033. Clark Univ. Press.

Young, A. B. F. and Smith, W. C. (1910–11). *Encyc. Brit.* 11th ed. **7**, 796.

Zahn, W. (1933). Ueber den Geruchsinn einiger Vögel. *Z. vergl. Physiol.* **19**, 785–96.

Zuckerman, S. (1932). *The Social Life of Monkeys and Apes.* London.

INDEX. PART I

BIRDS AND OTHER ORGANISMS

INDEX. PART II

SUBJECTS

INDEX. PART III

AUTHORS

Printed in the United States
By Bookmasters